Author

F. v. Haeseler
Katholieke Universiteit Leuven
Departement Elektrotechniek (ESAT/SISTA)
Kasteelpark Arenberg 10
3001 Leuven
Belgium

Mathematics Subject Classification 2000:
11Bxx, 68Rxx, 68Q

Key words:

Finite automata, automatic sequences, automatic maps, Mahler equations

Library of Congress − Cataloging-in-Publication Data

Haeseler, Friedrich von.
  Automatic sequences / by Friedrich von Haeseler.
    p.   cm − (De Gruyter expositions in mathematics; 36)
  Includes bibliographical references and index.
  ISBN 3 11 015629 6 (cloth : alk. paper)
    1. Sequences (Mathematics)   2. Algorithms.   I. Title.   II. Series.
QA292 .H34 2002
515′.24−dc21

2002031514

ISBN 3-11-015629-6

Bibliographic information published by Die Deutsche Bibliothek

Die Deutsche Bibliothek lists this publication in the Deutsche Nationalbibliografie;
detailed bibliographic data is available in the Internet at <http://dnb.ddb.de>.

Typesetting using the authors' TᴇX files: I. Zimmermann, Freiburg.
Printing and binding: Hubert & Co. GmbH & Co. KG, Göttingen.
Cover design: Thomas Bonnie, Hamburg.

Automatic Sequences

by

Friedrich von Haeseler

Walter de Gruyter · Berlin · New York 2003

de Gruyter Expositions in Mathematics

de Gruyter Expositions in Mathematics 36

Preface

This monograph is an extension of my Habilitationsschrift [95]. It is intended to generalize the 'classical' notion of automaticity to a more broader framework. The main purpose of this monograph is to demonstrate that the already existing 'classical' results can be rediscovered in the more general framework.

However, not all possible aspects of automatic sequences are covered. In particular, the connection between automatic sequences and logic is not even touched, see e.g. [47] and the literature cited there. Questions like complexity of sequences or spectral properties of sequences are also completely neglected, see e.g. [38], [39], [102], [133], [141], [142], [143].

Further relations to other structures are mentioned at the end of each chapter under the heading 'Notes and comments'.

Although it is not really necessary for an understanding of the basic ideas presented in the text, it is helpful if the reader is acquainted with finite fields, finitely generated groups, and some Perron–Frobenius Theory.

Acknowledgement. This monograph would have never been finished without the material support from the Universität Bremen, Florida Atlantic University (Boca Raton, U.S.A.), Katholieke Universiteit Leuven (Belgium).

Naturally moral support of every possible kind is much more important. It was provided by J. P. Allouche, A. Barbé, W. Jürgensen, M. Karbe, H.-O. Peitgen, A. Petersen, G. Skordev, and R. O. Wells. My sincere thanks to all of you.

Leuven, October 2002   F. v. Haeseler

Contents

Introduction

In recreational mathematics one often encounters the problem of continuing a given finite sequence of numbers, like,

$$1, \ 2, \ 3, \ 4, \ 5, \ldots$$

From a strictly logical point of view there does not exist a unique answer to questions of this type. In fact the answer can be whatever one likes. However, the 'real' meaning of the question is to find the rule or the algorithm (in a naive sense) which has produced the given finite sequence of numbers. Of course, as long as we are dealing with finite sequences, this question is meaningless, too. The situation changes if one considers infinite sequences, e.g., the decimal expansion of π,

$$\pi = 3.141592653589793238462\ldots$$

Then we all know that there exist several different algorithms to compute π. On the other hand, the naive approach of speaking of algorithms without a precise definition leads to severe problems. Tossing a coin infinitely often produces a sequence of 0's (tail) and 1's (head) in an algorithmic way. However, this 'algorithm' is hardly capable to produce the same sequence on a second run. One therefore is in need of a precise definition of an algorithm. Instead of troubling ourselves with the general definition of the notion of an algorithm, we simply drastically restrict the admissible rules to produce a sequence of numbers.

In this monograph we discuss sequences $(x_n)_{n \in \mathbb{N}}$ which are generated by a finite device. Loosely speaking, the value x_n can be determined in finitely many steps and from a knowledge of the p-adic expansion of n. Here the p-adic expansion of n is of the form

$$n = \sum_{j=0}^{\infty} n_j p^j,$$

where $p \geq 2$ is a natural number and $n_j \in \{0, 1, \ldots, p-1\}$. Sequences of this type will be called automatic sequences.

The goal of this introduction is twofold. Firstly, we provide a few examples which should give some insight into the realm of automatic sequences and their interrelations with other concepts. In the second part, we present a short summary of the results contained in this work.

We start with a classic example, the Thue–Morse sequence. In 1906, A. Thue constructed this sequence in connection with a word problem. In 1921, the Thue–Morse sequence was rediscovered by M. Morse to establish the existence of recurrent non-periodic geodesics. The Thue–Morse sequence $(t_n)_{n\in\mathbb{N}}$ is defined recursively by

$$t_0 = 0,$$
$$t_{2n} = t_n,$$
$$t_{2n+1} = 1 - t_n.$$

The first few terms are: $01101001100101\ldots$. The first property of the sequence (t_n) is a simple consequence of its definition. Let $2^l n_l + \cdots + 2 n_1 + n_0 = n$ be the 2-adic expansion of n, then

$$t_n = \left(\sum_{j=0}^{l} n_j\right) \bmod 2.$$

The next property of (t_n) is concerned with certain subsequences of (t_n). By construction of (t_n), we have $(t_{2n})_{n\in\mathbb{N}} = (t_n)_{n\in\mathbb{N}}$ and $(t_{2n+1})_{n\in\mathbb{N}} = (1 - t_n)_{n\in\mathbb{N}}$. Thus we can conclude that $t_{4n+1} = t_{2(2n)+1} = 1 - t_{2n} = 1 - t_n$ and $t_{4n+3} = t_{2(2n+1)+1} = 1 - t_{2n+1} = 1 - (1 - t_n) = t_n$. In other words, the set of subsequences which is defined by

$$\left\{ \left(t_{2^k n + j}\right)_{n\in\mathbb{N}} \mid k \in \mathbb{N}, \, j \in \{0, 1, 2, \ldots, 2^k - 1\}\right\}$$

is a finite set, namely $\{(t_n), (t_{2n+1})\}$. The above defined set of subsequences is called the 2-kernel of the Thue–Morse sequence. There exist certain relations between the elements of the 2-kernel of (t_n). We visualize this relations with a directed graph.

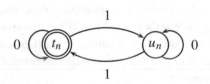

Figure 1. The graph associated with the Thue Morse sequence.

The interpretation of the graph is as follows: If one takes the subsequence (t_{2n}) of (t_n) one gets (t_n) back. If one takes the subsequence (t_{2n+1}) of (t_n) one gets a new sequence $(u_n) = (t_{2n+1})$. If one takes the subsequence (u_{2n}) of (u_n) one gets (u_n) back, for the other subsequence (u_{2n+1}) of (u_n) we obtain (t_n). Moreover, the above directed graph provides another method to compute t_n for a given n. If $2^l n_l + \cdots + 2 n_1 + n_0 = n$ denotes the 2-adic expansion of n, then the digits $n_0, n_1, \ldots, n_l$ define a path in the directed graph. The path starts at the vertex labelled (t_n) and then follows the arrows $n_0, n_1, \ldots, n_k$. The terminal vertex of the path determines t_n. If the path ends in (t_n) then we have $t_n = 0$, otherwise, we have $t_n = 1$. The graph is an example of a 2-automaton, and we say that the Thue–Morse sequence is 2-automatic, cf. Eilenberg [77], [78].

There is a third way to generate the Thue–Morse sequence. It is the method of substitution. A substitution is, roughly speaking, a rule which assigns words, i.e., concatenations of symbols or letters, to a single letter. The letters belong to a finite set which is called alphabet. We define a 2-substitution over the alphabet $\{0, 1\}$ by

$$0 \mapsto 01$$
$$1 \mapsto 10.$$

If we start with the letter 0 and iterate the substitution, we obtain a sequence of words

$$0 \mapsto 01 \mapsto 0110 \mapsto 01101001 \mapsto \cdots$$

and we observe that the sequence of words converges in a vague sense, to be more precise the beginning (if one reads from left to right) of the words tends to stabilize. In other words, we can speak of the limit sequence $(u_n)_{n \in \mathbb{N}}$ which is a fixed point of the substitution. By construction of the substitution, we have $u_{2n} = u_n$ and $u_{2n+1} = 1 - u_n$. Therefore, (u_n) is the Thue–Morse sequence. Thus, we can say that the Thue–Morse sequence is generated by a 2-substitution.

Besides this somehow combinatorial description of sequences which are generated by substitutions, there is yet another, more algebraic, characterization of automatic sequences. We explain the basic idea for the Thue–Morse sequence. Let (t_n) be the Thue–Morse sequence. We consider the formal power series

$$T(x) = \sum_{n=0}^{\infty} t_n x^n$$

as an element of $\mathbb{F}_2(x)$, i.e., the ring of all power series with coefficients in the field $\mathbb{F}_2 = \{0, 1\}$ with addition and multiplication modulo 2. Due to the characterization of the Thue–Morse sequence, we have the identities $t_{2n} = t_n$ and $t_{2n+1} = t_n + 1$, where the addition $t_n + 1$ has to be carried out in $\mathbb{Z}_2$. With these identities and the fact that for all $f(x) \in \mathbb{F}_2(x)$ we have $f(x)^2 = f(x^2)$ we obtain that $T(x)$ satisfies a Mahler equation, namely

$$
\begin{aligned}
T(x) &= \sum_{n=0}^{\infty} x^{2n} t_{2n} + \sum_{n=0}^{\infty} x^{2n+1} t_{2n+1} \\
&= \sum_{n=0}^{\infty} x^{2n} t_n + \sum_{n=0}^{\infty} x^{2n+1} (t_n + 1) \\
&= T(x)^2 + x T(x)^2 + x \sum_{n=0}^{\infty} x^{2n} \\
&= T(x)^2 (1 + x) + \frac{x}{1 + x^2}.
\end{aligned}
$$

As a summary we list the properties of the Thue–Morse sequence.

- The 2-kernel of (t_n) is finite.

- The Thue–Morse sequence is generated by a 2-automaton.

- There exists a 2-substitution which generates (t_n).

- The generating power series $T(x)$ of the Thue–Morse sequence satisfies an algebraic equation.

In [56], Cobham proved that the first three properties are equivalent. He showed that a sequence over a finite alphabet has a finite k-kernel if and only if the sequence is k-automatic, and the sequence is k-automatic if and only if it is generated by a k-substitution. The algebraic property of the Thue–Morse sequence reflects a celebrated theorem of Christol, Kamae, Mendès France, and Rauzy in [53], [54]. It states that a sequence (k_n) with values in a finite field $\mathbb{F}$ of prime characteristic p is p-automatic if the generating function, i.e., $f(x) = \sum k_n x^n$ satisfies a Mahler equation over the ring of formal power series $\mathbb{F}(x)$, that is

$$f(x) = \sum_{j=1}^{N} r_j(x) f(x^{p^j}),$$

where the $r_j(x)$ are rational functions. Furthermore, any solution in $\mathbb{F}(x)$ of the above equation is p-automatic.

As a further example and generalization, we study double sequences, i.e., $(s_{n,k})$, where $n, k \in \mathbb{N}$, with values in a finite set. One of the most prominent examples of such a double sequence are the binomial coefficients reduced modulo a natural number q, i.e., we consider $\left(\binom{n}{k} \bmod q\right)$, $k, n \in \mathbb{N}$. We start with the simplest example $(s_{k,n} = \binom{n}{k} \bmod 2)$. We consider the double sequence as a table of numbers. In the lower left corner we have $s_{0,0}$ and $s_{k,n}$ is the element in the k-th column (counted from the left) and in the n-th row (counted from the button). For the binomial coefficients we obtain the following

$$
\begin{array}{ccccccc}
\vdots & \vdots & \vdots & \vdots & \vdots & \vdots & \vdots \\
1 & 0 & 1 & 0 & 1 & 0 & 1 & \cdots \\
1 & 1 & 0 & 0 & 1 & 1 & 0 & \cdots \\
1 & 0 & 0 & 0 & 1 & 0 & 0 & \cdots \\
1 & 1 & 1 & 1 & 0 & 0 & 0 & \cdots \\
1 & 0 & 1 & 0 & 0 & 0 & 0 & \cdots \\
1 & 1 & 0 & 0 & 0 & 0 & 0 & \cdots \\
1 & 0 & 0 & 0 & 0 & 0 & 0 & \cdots
\end{array}
$$

By inspection, we conjecture that the above table is generated by the following 2-dimensional (2×2)-substitution:

$$
1 \mapsto \begin{array}{cc} 1 & 1 \\ 1 & 0 \end{array} \qquad 0 \mapsto \begin{array}{cc} 0 & 0 \\ 0 & 0 \end{array}.
$$

This fact can be proved using number theoretical properties of the binomial coefficients. As for the Thue–Morse sequence, we consider the formal power series

$$P(X, Y) = \sum_{k,n=0}^{\infty} s_{k,n} X^k Y^n,$$

where the $s_{k,n}$ are elements of the field $\mathbb{F}_2$. Due to the definition of the binomial coefficients, we have

$$P(X, Y) = \sum_{n=0}^{\infty} (1 + X)^n Y^n.$$

So, we have

$$P(X, Y) = \sum_{n=0}^{\infty} (1 + X)^{2n} Y^{2n} + \sum_{n=0}^{\infty} (1 + X)^{2n+1} Y^{2n+1}$$

$$= P(X^2, Y^2)(1 + Y + YX)$$

which yields

$$s_{2k,2n+1} = s_{k,n}, \quad s_{2k+1,2n+1} = s_{k,n}$$

$$s_{2k,2n} = s_{k,n}, \quad s_{2k+1,2n} = 0.$$

Actually, this is the substitution which we found by inspection of the table. On the other hand, we have $P(X, Y) = \frac{1}{1+Y(1+X)}$. Thus, we see that the binomial coefficients modulo 2 are generated by a 2-dimensional (2×2)-substitution and the generating power series satisfies an algebraic equation, $P(X, Y)(1 + Y(1 + X)) = 1$. It remains to show that $(s_{k,n})$ has a finite 2-dimensional (2×2)-kernel. By analogy, the (2×2)-kernel of a sequence $(u_{n,k})$ is defined by

$$\left\{ (u_{2^l k+j, 2^l n+i})_{k,n} \mid k \in \mathbb{N}, \ i, j \in \{0, 1, \ldots, 2^k - 1\} \right\}.$$

The above relations between $(s_{2k,2n})$, $(s_{2k+1,2n})$, $(s_{2k,2n+1})$, $(s_{2k+1,2n+1})$ imply that the (2×2)-kernel of the binomial coefficients mod 2 is given by $\{(s_{n,k}), (0)_{n,k}\}$, where $(0)_{n,k}$ is the zero sequence. The graph of the kernel relations is then given in Figure 2.

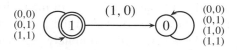

Figure 2. The graph associated with the binomial coefficients mod 2.

The graph is an example of a (2×2)-automaton which generates the binomial coefficients modulo 2. Let $(k, n) = 2^l (k_l, n_l) + 2^{l-1}(k_{l-1}, n_{l-1}) + \cdots + (k_0, n_0)$ be the 2-adic expansion of (k, n). As for the Thue–Morse sequence, the binary expansion of (k, n) defines a path which starts at the vertex 1. The terminal vertex determines $s_{k,n}$ which is 1 for the vertex 1 and 0 for the vertex 0.

For the general case, i.e., the binomial coefficients modulo q, one knows exactly whether or not the double sequence is automatic. In [11] and the literature cited there, the authors prove that the double sequence $(\binom{n}{k} \bmod q)$ is k-automatic if and only if k is a power of a prime number p and q is a power of p. In all other cases, there exists no k such that the double sequence is k-automatic.

There is yet another interpretation of the table of numbers which is expressed in [5], [6]. We can regard the double sequence as an orbit of a cellular automaton. We remind the reader what a cellular automaton is. Let A be a finite set and $\Sigma(\mathbb{Z}, A) = \{\underline{a} : \mathbb{Z} \to A\}$ be the set of bi-infinite sequences with values in A. A one-dimensional cellular automaton (of width r) is a map $C : \Sigma(\mathbb{Z}, A) \to \Sigma(\mathbb{Z}, A)$ such that

$$C(\underline{a})(i) = \Psi\big(\underline{a}(i - r), \underline{a}(i - r + 1), \ldots, \underline{a}(i), \ldots \underline{a}(i + r - 1), \underline{a}(i + r)\big),$$

where $\Psi : A^{2r+1} \to A$ is any map.

The above example, the binomial coefficients modulo 2, fits into the framework of cellular automata. Let $A = \mathbb{F}_2$ be the field with two elements. Then $\Sigma(\mathbb{Z}, \mathbb{F}_2)$ is the set of all Laurent series, i.e., $\underline{a} \in \Sigma(\mathbb{Z}, A)$ is written as $\underline{a} = \sum_{j=-\infty}^{\infty} a_j x^j$. The usual product of a polynomial with a Laurent series defines a cellular automaton. In our example, we have

$$(1 + x) \sum_{j=-\infty}^{\infty} a_j x^j = \sum_{j=-\infty}^{\infty} (a_{j-1} + a_j) x^j.$$

Thus the series $\sum_{n=0}^{\infty} Y^n (1 + X)^n$ describes the orbit of the Laurent series 1 under iteration of the cellular automata. Orbits of certain cellular automata provide examples of automatic double sequences.

The goal of this work is to develop a general approach to the above presented phenomena.

In Chapter 1, we introduce the general setting. The set $\mathbb{N}$ (for the Thue–Morse sequence) or the set $\mathbb{N}^2$ (for the binomial coefficients modulo 2) is replaced by a finitely generated group Γ. By abuse of language, we consider 'sequences' to be maps from Γ to a finite set $\mathcal{A}$. The set of sequences is denoted by $\Sigma(\Gamma, \mathcal{A})$.

The second chapter deals with the introduction of the notion of a substitution on $\Sigma(\Gamma, \mathcal{A})$. In the examples above we considered the p-adic expansion of a natural number. This concept can be generalized to a finitely generated group Γ, provided there exists an expanding group endomorphism $H : \Gamma \to \Gamma$ such that the subgroup $H(\Gamma)$ is of finite index. The first section discusses under which conditions on the group one can find such an expanding endomorphism.

In the second part of Chapter 2 we introduce the general notion of a substitution $S : \Sigma(\Gamma, \mathcal{A}) \to \Sigma(\Gamma, \mathcal{A})$. More precisely, a (V, H)-substitution, where $H : \Gamma \to \Gamma$

is an expanding group endomorphism with $H(\Gamma)$ of finite index and V is a complete system of left representatives of $\Gamma/H(\Gamma)$.

As for the above examples, we study the fixed points of a substitution S. Analogous to our examples, we introduce the (V, H)-kernel of a sequence. The main result will be: A sequence $\underline{f} \in \Sigma(\Gamma, \mathcal{A})$ is generated by a (V, H)-substitution if and only if it has a finite (V, H)-kernel. In this chapter we also introduce the kernel graph of a sequence. The kernel graph will be an important tool in the later chapters.

Chapter 3 is devoted to the introduction of (V, H)-automata, a notion analogous to the finite automata given above. A (V, H)-automatic sequence is generated by a finite (V, H)-automaton. We will prove that the following three statements are equivalent:

- $\underline{f}$ is generated by a finite (V, H)-automaton,

- $\underline{f}$ is generated by a (V, H)-substitution,

- $\underline{f}$ has a finite (V, H)-kernel.

In the second part of Chapter 3 a further investigation of (V, H)-automatic sequences leads to the result that the above three statements are independent of the set V. E.g., if V and W are different systems of left representatives, then the (V, H)-kernel of $\underline{f}$ is finite if and only if the (W, H)-kernel of $\underline{f}$ is finite. It is therefore justified to speak of H-automatic sequences rather then of (V, H)-automatic sequences.

Chapter 4 pursues the investigation of properties of H-automatic sequences. In the first section we demonstrate that a study of H-automatic sequences can be reduced to a study of H-automatic subsets of Γ. We introduce the notion of $(H_1 \times H_2)$-automatic maps $G : \Gamma_1 \to \Gamma_2$ and study their properties. The most important property is that they preserve automaticity, i.e., if G is an $(H_1 \times H_2)$-automatic map, then G maps an H_1-automatic subset on an H_2-automatic subset. In the third part of Chapter 4, we characterize $(H_1 \times H_2)$-automata which generate an $(H_1 \times H_2)$-automatic map. Furthermore, we prove that there exists a universal map $\mathfrak{V}_p : \Gamma \to \mathbb{N}$ such that for every H-automatic subset $M \subset \Gamma$ the image $\mathfrak{V}_p(M)$ is a p-automatic subset of $\mathbb{N}$, where $p = |V|$ is the cardinality of V.

In the fourth section automatic functions $G : \mathbb{N} \to \mathbb{N}$ are studied in detail. We present upper and lower estimates on the growth of these functions and we discuss the question on the existence of automatic functions with prescribed properties. E.g., we show that for $p < q$ no surjective (p, q)-automatic function exists.

Chapter 5 is written in the spirit of the above mentioned results on Mahler equations. We introduce an additional algebraic structure on the finite set $\mathcal{A}$. Firstly, we suppose that $\mathcal{A}$ is a finite, commutative monoid. Closely related to the monoid is the set of endomorphisms of $\mathcal{A}$, i.e., the set $\mathrm{End}(\mathcal{A}) = \{h : \mathcal{A} \to \mathcal{A} \mid h(a + b) = h(a) + h(b)$ holds for all $a, b \in \mathcal{A}\}$ and $+$ denotes the addition in the monoid. A polynomial with coefficients in $\mathrm{End}(\mathcal{A})$ is a sequence (h_γ) over Γ such that $h_\gamma = 0$ (the zero-map) almost everywhere. The set of polynomials operates on the set of sequences over Γ with values in $\mathcal{A}$. The important result of Chapter 5 is that any solution $\underline{f} = (f_\gamma)_{\gamma \in \Gamma}$ of a Mahler equation of the type

$$\underline{f} = p_1 \cdot H_*(\underline{f}) + \cdots + p_N \cdot H_*^{\circ N}(\underline{f})$$

is generated by a substitution and therefore H-automatic. In the above formula, p_j are polynomials with coefficients in $\mathrm{End}(\overline{\mathcal{A}})$, $\cdot$ is the operation of polynomials on sequences, and H_* is analogous to the map $f(x) \mapsto f(x^2)$ for the above mentioned Thue–Morse sequence.

In the second part of Chapter 5, we present two methods how to solve a Mahler equation of the form

$$\underline{f} = p \cdot H_*(\underline{f}).$$

One method is based on the construction of a certain substitution, the other one is based on the computation of the kernel graph of an assumed solution.

In the last section of Chapter 5 we consider sequences over Γ, where Γ is commutative and the sequences take their values in a finite field. Under these strong conditions we will be able to establish the existence of certain Mahler equations for automatic sequences.

Chapter 1

Preliminaries

In this chapter, we present the basic notions which will occur throughout the course of this book. We begin with some basic properties of norms on groups and continue with the basic facts on sequence spaces.

In the second part of this chapter, we study a particular group, the Heisenberg group. It serves as an important example of a non-Abelian group for our theory.

1.1 Sequence spaces over groups

Let Γ be a group and denote by e the unit of the group Γ, i.e., $e\gamma = \gamma e = \gamma$ for all $\gamma \in \Gamma$. A norm on Γ is a function $\| \, \| : \Gamma \to \mathbb{R}_0^+ = \{r \mid r \in \mathbb{R}, r \geq 0\}$ with the following properties:

a) $\|\gamma\| = 0$ if and only if $\gamma = e$

b) $\|\gamma\| = \|\gamma^{-1}\|$ for all $\gamma \in \Gamma$

c) $\|\gamma'\gamma\| \leq \|\gamma'\| + \|\gamma\|$ for all $\gamma', \gamma \in \Gamma$.

If Γ is a finitely generated group, e.g., generated by the set $E = \{\gamma_1, \ldots, \gamma_k\}$, then there exists at least one function having the properties a), b), and c). Each element $\gamma \in \Gamma$ can be represented by a word $\gamma_{i_1}^{p_1} \gamma_{i_2}^{p_2} \ldots \gamma_{i_l}^{p_l}$, with $p_j \in \mathbb{Z}$ for $j = 1, \ldots, l$. The number $|p_1| + \cdots + |p_l|$ is called the length of the word [1]. The length of the identity element is equal to zero. We define the function $\|\gamma\|_E$ (with respect to $E = \{\gamma_1, \ldots, \gamma_k\}$) as the minimal length of the word representing γ. Then the function $\| \, \|_E : \Gamma \to \mathbb{R}$ has all the properties of a norm. We call $\| \, \|_E$ the norm induced by E. Moreover, if $E' = \{\delta_1, \ldots, \delta_l\}$ is another set of generators of Γ and $\| \, \|_{E'}$ the induced norm, then there exists a constant $C > 0$ such that

$$C\|\gamma\|_E \geq \|\gamma\|_{E'} \geq C^{-1}\|\gamma\|_E$$

holds for all $\gamma \in \Gamma$.

---

[1] Sometimes we define the length of a word as $\sqrt{p_1^2 + \cdots + p_l^2}$ or any other norm of this type. In any case, every such word length gives a norm on Γ.

Any norm $\| \ \| : \Gamma \to \mathbb{R}_0^+$ on Γ induces a metric $D_{\| \ \|}$ on Γ. We define $D_{\| \ \|} : \Gamma \times \Gamma \to \mathbb{R}_0^+$ by

$$D_{\| \ \|}(\gamma, \rho) = \|\rho^{-1}\gamma\|.$$

By construction, the metric $D_{\| \ \|}$ is invariant under left translations, i.e., we have

$$D_{\| \ \|}(\tau\gamma, \tau\rho) = D_{\| \ \|}(\gamma, \rho)$$

for all $\tau \in \Gamma$. On the other hand, any left invariant metric $D : \Gamma \times \Gamma \to \mathbb{R}_0^+$ induces a norm on Γ by setting $\|\gamma\|_D = D(\gamma, e)$.

From now on we suppose that Γ is equipped with a norm which we denote by $\| \ \|$. In most cases we may think of the norm as being induced from a set of generators of the group.

It makes perfect sense to speak of balls of the group. The subset $B_r(\gamma) \in \Gamma$ defined by

$$B_r(\gamma) = \{\tau \mid \tau \in \Gamma \text{ and } \|\gamma^{-1}\tau\| \le r\}$$

is called the *closed ball of radius r with center γ*. The *open* ball of radius r with center in γ, denoted by $\overset{o}{B}_r(\gamma)$, is the set of $\tau \in \Gamma$ with $\|\gamma^{-1}\gamma'\| < r$. If we want to emphasize the norm for which we consider the balls, we use the precise language $\| \ \|$-ball.

Examples.

1. Let $\mathbb{Z}$ be the group of integers with addition. Then $\mathbb{Z}$ is generated by 1, and for $l \in \mathbb{Z}$ we have for the induced norm $\|l\| = |l|$, where $|\ |$ is the absolute value function.

2. Since 2 and 3 are relatively prime, any integer $l \in \mathbb{Z}$ can be written as $l = 2m + 3n$, with integers $m, n \in \mathbb{Z}$. Note that the representation of l is not unique. Therefore 2 and 3 generate $\mathbb{Z}$, and the induced norm is given by

$$\|l\| = \begin{cases} k & \text{if } |l| = 3k \\ k+2 & \text{if } |l| = 3k+1 \\ k+1 & \text{if } |l| = 3k+2. \end{cases}$$

3. Let Γ be the free group generated by a, b, then the induced norm of an element $\gamma \in \Gamma$ is the length of γ, i.e., if $\gamma = \alpha_1\alpha_2 \ldots \alpha_n$, with $\alpha_i \in \{a, a^{-1}, b, b^{-1}\}$, $i = 1, \ldots, n$, then $\|\gamma\| = n$.

4. Let $\mathscr{G} = \mathbb{Z} + \imath\mathbb{Z}$ be the additive group of the Gaussian integers. The group $\mathscr{G}$ is generated by $\{1, \imath\}$ and the induced norm of $\omega \in \mathscr{G}$ is given by $\|\omega\| = |\Re(\omega)| + |\Im(\omega)|$, where $\Re(\omega)$ denotes the real part of ω and $\Im(\omega)$ denotes the imaginary part of ω.

5. Let m be a natural number greater than or equal to 2 and let $G_m = \{0, 1, \ldots, m-1\} = \mathbb{Z}/m\mathbb{Z}$ denote the additive group of residue classes modulo m. Then G_m is generated by 1 and the induced norm $\| \ \|_1$ of an element $g \in G_m$ equals g. If h is a generator of G_m then the induced norm $\| \ \|_h$ of an element g is given by $\|g\|_h = hg \bmod m$.

Let Γ_1 and Γ_2 be finitely generated groups with norms $\| \ \|_1$ and $\| \ \|_2$. Then the direct product $\Gamma_1 \times \Gamma_2$ is a finitely generated group and the function $\| \ \| : \Gamma_1 \times \Gamma_2 \to \mathbb{R}_0^+$ defined by $\|(\gamma, \rho)\| = \|\gamma\|_1 + \|\rho\|_2$ is a norm for the direct product. In the same way we can define a norm on the n-fold direct product of groups.

We now introduce the main object of our interest. It is the notion of a sequence over a group Γ, or, if is clear from the context, simply a sequence. In general, we assume that a sequence has its values in a finite set.

Let $\mathcal{A} = \{a_1, \ldots, a_N\}$ be a finite set, where $N \in \mathbb{N}$, and denote by $\Sigma(\Gamma, \mathcal{A}) = \mathcal{A}^\Gamma = \{\underline{f} : \Gamma \to \mathcal{A}\}$ the set of maps from Γ to $\mathcal{A}$. The elements of $\mathcal{A}$ are called symbols and the elements of $\Sigma(\Gamma, \mathcal{A})$ are called *sequences* over Γ with values in the symbols $\mathcal{A}$.

Any sequence space can be equipped with a metric in such a way that it becomes a complete metric space. To this end, let δ be any metric on $\mathcal{A}$ and let $(r_\gamma)_{\gamma \in \Gamma}$ be a family of positive numbers such that $\sum_{\gamma \in \Gamma} r_\gamma < \infty$. Then

$$\Delta(\underline{f}, \underline{g}) = \sum_{\gamma \in \Gamma} r_\gamma \delta(\underline{f}(\gamma), \underline{g}(\gamma)) \tag{1.1}$$

defines a metric on $\Sigma(\Gamma, \mathcal{A})$ such that $(\Sigma(\Gamma, \mathcal{A}), \Delta)$ is a compact metric space. In fact, if the cardinality of $\mathcal{A}$ is greater than 1, then $(\Sigma(\Gamma, \mathcal{A}), \Delta)$ is a Cantor set. From now on, $\Sigma(\Gamma, \mathcal{A})$ will be considered as a metric space with a metric of the above type.

For several technical reasons which will become apparent later, we have to enlarge the set of symbols $\mathcal{A}$ by an *empty symbol*, denoted by $\varnothing$ and which is not an element of $\mathcal{A}$. If the set of symbols is enlarged by the empty symbol then we denote it by $\overline{\mathcal{A}}$. We are mainly interested in sequences with values in the symbol set $\mathcal{A}$. Sequences with values in the extended symbol set $\overline{\mathcal{A}}$ are only a helpful tool in certain constructions. In all of the following we consider $\Sigma(\Gamma, \mathcal{A})$ as a subset of $\Sigma(\Gamma, \overline{\mathcal{A}})$.

If we consider sequences with values in a product of finite sets, i.e., a set of the form $\mathcal{A}_1 \times \mathcal{A}_2 \times \cdots \times \mathcal{A}_n$. Then $\overline{\mathcal{A}_1 \times \mathcal{A}_2 \times \cdots \times \mathcal{A}_n}$ denotes the extension of the product by the empty symbol.

The sequence $\underline{\varnothing}(\gamma) \equiv \varnothing$ is called the *empty sequence*.

Let $\mathcal{A}$, $\mathcal{B}$ be two finite sets and $a : \mathcal{A} \to \mathcal{B}$ be a map, then a has the obvious extension, also denoted by a, $a : \overline{\mathcal{A}} \to \overline{\mathcal{B}}$, where $a(\varnothing) = \varnothing$.

For $\underline{f} \in \Sigma(\Gamma, \overline{\mathcal{A}})$, we denote by $\mathrm{supp}(\underline{f}) = \{\gamma \mid \underline{f}(\gamma) \neq \varnothing\}$ the support of $\underline{f}$. The set of sequences with finite support is denoted by $\Sigma_c(\Gamma, \overline{\mathcal{A}})$. By abuse of language, elements of $\Sigma_c(\Gamma, \overline{\mathcal{A}})$ are called *polynomials with coefficients* in $\overline{\mathcal{A}}$.

For $\underline{f}, \underline{g} \in \Sigma(\Gamma, \overline{\mathcal{A}})$ with $\operatorname{supp}(\underline{f}) \cap \operatorname{supp}(\underline{g}) = \emptyset$, we define the sum $\underline{f} \oplus \underline{g}$ of $\underline{f}$ and $\underline{g}$ by

$$(\underline{f} \oplus \underline{g})(\gamma) = \begin{cases} \underline{f}(\gamma) & \text{if } \gamma \in \operatorname{supp}(\underline{f}) \\ \underline{g}(\gamma) & \text{if } \gamma \in \operatorname{supp}(\underline{g}) \\ \emptyset & \text{otherwise.} \end{cases} \qquad (1.2)$$

For $\tau \in \Gamma$ and $a \in \mathcal{A}$, we introduce the special sequence $\underline{f}_\tau^a : \Gamma \to \overline{\mathcal{A}}$ having the property that

$$\underline{f}_\tau^a(\gamma) = \begin{cases} a & \text{if } \gamma = \tau \\ \emptyset & \text{otherwise.} \end{cases}$$

In this notation, we can write any $\underline{f} \in \Sigma(\Gamma, \overline{\mathcal{A}})$ as a sum

$$\underline{f} = \bigoplus_{\gamma \in \operatorname{supp}(\underline{f})} \underline{f}_\gamma^{f(\gamma)}.$$

For the sake of simplicity and for later use, we abbreviate the above notation as a "formal series", i.e., we write $\underline{f}$ as

$$\underline{f} = \bigoplus_{\gamma \in \operatorname{supp}(\underline{f})} \gamma f_\gamma,$$

where $f_\gamma = \underline{f}(\gamma) \in \mathcal{A}$. If it is clear from the context, we even omit the summation index.

Then the sum of $\underline{f}$ and $\underline{g}$, as defined in Equation (1.2), can be written as

$$\underline{f} \oplus \underline{g} = \bigoplus \gamma f_\gamma \oplus \bigoplus \gamma g_\gamma = \bigoplus \gamma (f_\gamma \oplus g_\gamma),$$

where $\oplus$ is defined in the obvious way.

As a next step, we introduce the notion of reduction and expansion maps. These maps are of utmost importance for all of the following.

Definition 1.1.1. Let $G : \Gamma \to \Gamma$ be an arbitrary map. Then G induces a map $G^* : \Sigma(\Gamma, \overline{\mathcal{A}}) \to \Sigma(\Gamma, \overline{\mathcal{A}})$ which is defined by

$$G^*(\underline{f})(\gamma) = \underline{f} \circ G(\gamma).$$

G^* is called the *G-reduction*.

Furthermore, for G injective there exists a second map $G_* : \Sigma(\Gamma, \overline{\mathcal{A}}) \to \Sigma(\Gamma, \overline{\mathcal{A}})$ which is defined by

$$G_*(\underline{f})(\gamma) = \begin{cases} \emptyset & \text{if } \gamma \notin G(\Gamma) \\ \underline{f}(\rho) & \text{if } \gamma = G(\rho). \end{cases}$$

The map G_* is called the *G-expansion*.

Remark. The set $\Sigma(\Gamma, \mathcal{A})$ is invariant under G-reductions; but $\Sigma(\Gamma, \mathcal{A})$ is not invariant under G-expansions. Note that $\Sigma(\Gamma, \overline{\mathcal{A}})$ is invariant under G-expansions.

Examples.

1. Let $\mathcal{A}$ be a finite set and let $\Gamma = \mathbb{Z}$ be the additive group of natural numbers, let $G : \mathbb{Z} \to \mathbb{Z}$ be defined by $G(z) = 3z + 1$. Then the G-reduction of a sequence $\underline{f} = \oplus j f_j$ is given by

$$G^*(\underline{f}) = \bigoplus j f_{3j+1}.$$

If we consider the sequence $\underline{f}$ in the following way

$$\underline{f} = (\ldots \quad f_{-4} \quad f_{-3} \quad f_{-2} \quad f_{-1} \quad f_0 \quad f_1 \quad f_2 \quad f_3 \quad f_4 \quad \ldots),$$
$$\uparrow$$

where the arrow indicates the zero position, then $G^*(\underline{f})$ is the sequence given by

$$G^*(\underline{f}) = (\ldots \quad f_{-11} \quad f_{-8} \quad f_{-5} \quad f_{-2} \quad f_1 \quad f_4 \quad f_7 \quad f_{10} \quad f_4 \ldots).$$
$$\uparrow$$

This explains the name reduction.

The G-expansion of a sequence $\underline{f}$ is given by

$$G_*(\underline{f}) = \bigoplus (3j + 1) f_j,$$

which can be regarded as the sequence given by

$$G_*(\underline{f}) = (\ldots \quad \varnothing \quad \varnothing \quad f_{-1} \quad \varnothing \quad \varnothing \quad f_0 \quad \varnothing \quad \varnothing \quad f_2 \ldots).$$
$$\uparrow$$

In other words, a G-expansion expands a sequence with values in $\mathcal{A}$ (!) by inserting the empty symbol at certain places.

2. Let $G : \mathbb{Z} \to \mathbb{Z}$ be defined by

$$G(z) = \begin{cases} 1 & \text{if } z = 0 \\ 2 & \text{if } z = 1 \\ 0 & \text{if } z = 2 \\ z & \text{otherwise,} \end{cases}$$

then $G^*(\underline{f})$ is given by

$$\underline{f} = (\ldots \quad f_{-4} \quad f_{-3} \quad f_{-2} \quad f_{-1} \quad f_1 \quad f_2 \quad f_0 \quad f_3 \quad f_4 \quad \ldots)$$
$$\uparrow$$

and $G_*(\underline{f})$ is given by

$$\underline{f} = (\ldots \quad f_{-4} \quad f_{-3} \quad f_{-2} \quad f_{-1} \quad f_2 \quad f_0 \quad f_1 \quad f_3 \quad f_4 \quad \ldots).$$
$$\uparrow$$

The example shows that for a map $G : \Gamma \to \Gamma$ being surjective the G-expansion is well defined on $\Sigma(\Gamma, \mathcal{A})$.

Remarks.

1. We have the obvious relations for maps G, H

$$(G \circ H)^* = H^* \circ G^*,$$

and for G, H injective we have

$$(G \circ H)_* = G_* \circ H_*.$$

2. In the notation of formal series, we have

$$G^*(\underline{f}) = \bigoplus \gamma f_{G(\gamma)},$$

and for G injective

$$G_*(\underline{f}) = \bigoplus G(\gamma) f_\gamma.$$

Proposition 1.1.2. *Let* $G : \Gamma \to \Gamma$ *be an injective map. Then*

$$G^* \circ G_* = \mathrm{id},$$

where id *denotes the identity on* $\Sigma(\Gamma, \overline{\mathcal{A}})$.

Proof. We have

$$G^*(G_*(\underline{f}))(\gamma) = G_*(\underline{f})(G(\gamma)) = \underline{f}(\gamma)$$

due to Definition 1.1.1.   □

In other words, G^* is the left inverse of G_*. Moreover, note that for $G : \Gamma \to \Gamma$ bijective, the maps G^* and G_* are well defined on $\Sigma(\Gamma, \mathcal{A})$, and we have that $G_* = (G^{-1})^*$, or, equivalently, $(G^{-1})_* = G^*$.

Given two finite sets $\mathcal{A}$ and $\mathcal{B}$ and a map $a : \mathcal{A} \to \mathcal{B}$. Then the map a induces a map $\hat{a} : \Sigma(\Gamma, \overline{\mathcal{A}}) \to \Sigma(\Gamma, \overline{\mathcal{B}})$ which is defined by

$$\hat{a}(\underline{f})(\gamma) = a(\underline{f}(\gamma))$$

for all $\gamma \in \Gamma$. In the notation of formal series we write

$$\hat{a}(\underline{f}) = \bigoplus \gamma a(f_\gamma).$$

Proposition 1.1.3. *Let* $G : \Gamma \to \Gamma$ *and* $a : \mathcal{A} \to \mathcal{A}$ *be maps, then*

$$\hat{a} \circ G^* = G^* \circ \hat{a}.$$

If, additionally, G *is injective, then*

$$\hat{a} \circ G_* = G_* \circ \hat{a}.$$

The proof is an immediate consequence of the definitions.

Let τ be an element of Γ, then the map

$$T_\tau : \Gamma \to \Gamma$$
$$\gamma \mapsto \tau\gamma \qquad\qquad (1.3)$$

is called *left translation*. The induced map $T_\tau^* : \Sigma(\Gamma, \overline{\mathcal{A}}) \to \Sigma(\Gamma, \overline{\mathcal{A}})$ is called a (*left*) *shift* on the set of sequences. Moreover, since T_τ is bijective, we have $T_\tau^* = (T_{\tau^{-1}})_*$ and $\Sigma(\Gamma, \mathcal{A})$ is invariant under T_τ^* and $(T_\tau)_*$.

Using the notion of formal series we have

$$(T_\tau)_*(\underline{f}) = \oplus\tau\gamma f_\gamma$$

and we use the short form $\tau\underline{f}$ for $(T_\tau)_*(\underline{f})$. The *right* translations are defined in a similar way, namely

$$R_\tau : \Gamma \to \Gamma$$
$$\gamma \mapsto \gamma\tau. \qquad\qquad (1.4)$$

The induced map R_τ^* on $\Sigma(\Gamma, \overline{\mathcal{A}})$ is called a right shift.

1.2   The Heisenberg group

In this section, we introduce one of our main examples of a non-commutative group to which all of our theory can be applied. It is a finitely generated discrete subgroup of the so-called Heisenberg group. The Heisenberg group is a non-Abelian, simply connected 3-dimensional Lie group. Following the definition of the Heisenberg group we show how to equip the Heisenberg group with a left invariant Riemannian metric w.r.t. left translations. This left invariant metric gives a norm on the discrete subgroups of the Heisenberg group.

Let L denote the set of upper triangular matrices with 1's on the diagonal, i.e.,

$$L = \left\{ \begin{pmatrix} 1 & x & z \\ 0 & 1 & y \\ 0 & 0 & 1 \end{pmatrix} \;\middle|\; x, y, z \in \mathbb{R} \right\}.$$

The usual matrix multiplication provides a group structure, where the unit is given by the unit matrix. Moreover, the multiplication is smooth. If

$$A = \begin{pmatrix} 1 & x & z \\ 0 & 1 & y \\ 0 & 0 & 1 \end{pmatrix}$$

is in L, then the inverse A^{-1} of A is given by

$$A^{-1} = \begin{pmatrix} 1 & -x & xy - z \\ 0 & 1 & -y \\ 0 & 0 & 1 \end{pmatrix}.$$

and therefore a smooth map. Thus, L is a Lie group. Moreover, we see that the set

$$L_{\mathbb{Z}} = \left\{ \begin{pmatrix} 1 & x & z \\ 0 & 1 & y \\ 0 & 0 & 1 \end{pmatrix} \;\middle|\; x, y, z \in \mathbb{Z} \right\},$$

is a discrete subgroup of L.

Let $\Gamma_H = \langle a, b, c; abc = ba, ac = ca, bc = cb \rangle$, i.e., Γ_H is generated as a group by a, b, c with relations $abc = ba$, $ac = ca$, and $bc = cb$. The group Γ_H is called the discrete Heisenberg group.

Lemma 1.2.1.

1. $a^m b^n = b^n a^m c^{-nm}$ for all $n, m \in \mathbb{Z}$.

2. $(b^n a^m c^p)^{-1} = b^{-n} a^{-m} c^{-p-nm}$ for all $n, mp \in \mathbb{Z}$.

3. The group Γ_H is isomorphic to $L_{\mathbb{Z}}$.

Proof. 1. We have $abc = ba$ which gives $ab = bac^{-1}$ and $a^{-1} b^{-1} = b^{-1} a^{-1} c^{-1}$. From $abc = ba$ we obtain $b = a^{-1} bac^{-1}$ and thus $ab^{-1} = a(a^{-1} b^{-1} ac) = b^{-1} ac$ and $a^{-1} b = ba^{-1} c$. This gives $a^\epsilon b^\delta = b^\delta a^\epsilon c^{-\epsilon\delta}$, for $\epsilon, \delta \in \{-1, 0, 1\}$.

If we define $\epsilon : \mathbb{Z} \to \{-1, 0, 1\}$ by

$$\epsilon(k) = \begin{cases} -1 & \text{if } k \leq -1 \\ 0 & \text{if } k = 0 \\ 1 & \text{if } k \geq 1, \end{cases}$$

then $a^m b^n$ can be written as

$$a^m b^n = \underbrace{a^{\epsilon(m)} \ldots a^{\epsilon(m)}}_{|m|\text{-times}} \underbrace{b^{\epsilon(n)} \ldots b^{\epsilon(n)}}_{|n|\text{-times}},$$

and an nm-times application of $a^\epsilon b^\delta = b^\delta a^\epsilon c^{-\epsilon\delta}$ yields the desired result.

2. One has $(b^n a^m c^p)^{-1} = c^{-p} a^{-m} b^{-n} = a^{-m} b^{-n} c^{-p}$ and by 1. we obtain the assertion.

3. Let $\gamma \in \Gamma_H$, then $\gamma = \gamma_1^{\alpha_1} \ldots \gamma_k^{\alpha_k}$ for some k, $\gamma_j \in \{a, b, c\}$, $j = 1, \ldots, k$ and $\alpha_1, \ldots, \alpha_k \in \mathbb{Z}$. By the defining relations for Γ_H, we can write γ as

$$\gamma = \gamma_1^{\alpha_1} \ldots \gamma_n^{\alpha_n} c^{p'},$$

where $\gamma_j \in \{a, b\}$, $j = 1, \ldots, k$. By 1., we see that we can move all a's to the right adding or removing c's. After finitely many steps we have that γ is of the form

$$\gamma = b^n a^m c^p.$$

This gives a map $h : \Gamma_H \to L_{\mathbb{Z}}$ defined by

$$h(b^n a^m c^p) = \begin{pmatrix} 1 & n & p \\ 0 & 1 & m \\ 0 & 0 & 1 \end{pmatrix}.$$

By construction of h, we have the identity

$$h(b^n a^m c^p)\, h(b^{n'} a^{m'} c^{p'}) = h(b^n a^m c^p\, b^{n'} a^{m'} c^{p'}) = h(a^{n+n'} b^{m+m'} c^{p+p'+nm'}),$$

which shows that h is a morphism of groups. Moreover, h is a bijection, i.e., both groups are isomorphic.   □

As the proof of Lemma 1.2.1 shows, $L_{\mathbb{Z}}$ is generated by the matrices

$$A = \begin{pmatrix} 1 & 0 & 0 \\ 0 & 1 & 1 \\ 0 & 0 & 1 \end{pmatrix}, \quad B = \begin{pmatrix} 1 & 1 & 0 \\ 0 & 1 & 0 \\ 0 & 0 & 1 \end{pmatrix}, \quad C = \begin{pmatrix} 1 & 0 & 1 \\ 0 & 1 & 0 \\ 0 & 0 & 1 \end{pmatrix},$$

and we have $ABC = BA$, $AC = CA$, $BC = CB$.

As a next step, we define a left invariant Riemannian metric on L. Let

$$\mathcal{L} = \left\{ \begin{pmatrix} 0 & x & z \\ 0 & 0 & y \\ 0 & 0 & 0 \end{pmatrix} \,\middle|\, x, y, z \in \mathbb{R} \right\}$$

denote the nilpotent Lie algebra of L. On $\mathcal{L}$ we introduce a norm by defining $\|a\| = \sqrt{x^2 + y^2 + z^2}$, where $a \in \mathcal{L}$. We now consider the Heisenberg group as a 3-dimensional manifold with tangent space isomorphic to $\mathcal{L}$ at each point of the Heisenberg group. Let

$$v = \begin{pmatrix} 0 & \xi & \zeta \\ 0 & 0 & \eta \\ 0 & 0 & 0 \end{pmatrix} \in T_{\begin{pmatrix} 1 & x & z \\ 0 & 1 & y \\ 0 & 0 & 1 \end{pmatrix}} L$$

be a tangent vector, then we define the norm by $\|v\| = \sqrt{(\zeta - x\eta)^2 + \eta^2 + \xi^2}$. This gives a Riemannian metric on L which is invariant under left translations. To this end, let

$$\begin{pmatrix} 1 & \alpha & \gamma \\ 0 & 1 & \beta \\ 0 & 0 & 1 \end{pmatrix} \in L$$

and denote by $T_{\alpha,\beta,\gamma} : L \to L$ the associated left translation, i.e., the left multiplication of the above matrix with elements of the Heisenberg group. Then the derivative $DT_{\alpha,\beta,\gamma}$ of $T_{\alpha,\beta,\gamma}$, viewed as a linear map from $\mathcal{L}$ to $\mathcal{L}$, is given by

$$DT_{\alpha,\beta,\gamma} : \mathcal{L} \to \mathcal{L}$$

$$\begin{pmatrix} 0 & \xi & \zeta \\ 0 & 0 & \eta \\ 0 & 0 & 0 \end{pmatrix} \mapsto \begin{pmatrix} 1 & \alpha & 0 \\ 0 & 1 & 0 \\ 0 & 0 & 1 \end{pmatrix} \begin{pmatrix} 0 & \xi & \zeta \\ 0 & 0 & \eta \\ 0 & 0 & 0 \end{pmatrix},$$

Therefore $DT_{\alpha,\beta,\gamma}$ is a map from

$$DT_{\alpha,\beta,\gamma} : T_{\left(\begin{smallmatrix} 1 & x & z \\ 0 & 1 & y \\ 0 & 0 & 1 \end{smallmatrix}\right)} L \to T_{\left(\begin{smallmatrix} 1 & x+\alpha z+\gamma+y\alpha \\ 0 & 1 & y+\beta \\ 0 & 0 & 1 \end{smallmatrix}\right)} L$$

and for

$$v = \begin{pmatrix} 0 & \xi & \zeta \\ 0 & 0 & \eta \\ 0 & 0 & 0 \end{pmatrix} \in T_{\left(\begin{smallmatrix} 1 & x & z \\ 0 & 1 & y \\ 0 & 0 & 1 \end{smallmatrix}\right)} L$$

we compute $DT_{\alpha,\beta,\gamma} v$ as

$$DT_{\alpha,\beta,\gamma} v = \begin{pmatrix} 0 & \xi & \zeta+\alpha\eta \\ 0 & 0 & \eta \\ 0 & 0 & 0 \end{pmatrix} \in T_{\left(\begin{smallmatrix} 1 & x+\alpha z+\gamma+y\alpha \\ 0 & 1 & y+\beta \\ 0 & 0 & 1 \end{smallmatrix}\right)} L.$$

The norm of $DT_{\alpha,\beta,\gamma} v$ is given by

$$\|DT_{\alpha,\beta,\gamma} v\| = (\zeta + \alpha\eta - (x+\alpha)\eta)^2 + \eta^2 + \xi^2$$

which is the same as $\|v\|$. If we define the distance of two points as the length (w.r.t. the metric defined above) of the shortest path connecting these two points, we obtain a left invariant metric on L. Therefore we have a left invariant metric on the subgroup $L_{\mathbb{Z}}$ which yields a left invariant metric on Γ_H.

From now on we suppose that Γ_H is equipped with the norm which is induced by the Riemannian metric defined on L, and we denote this norm by $\| \ \|_r$.

1.3   Notes and comments

The definition of the norm $\| \ \|_E$ for a finitely generated group is from [88]. A proof of the fact that the metric space $(\Sigma(\Gamma, \mathcal{A}), \Delta))$ is a Cantor set can be found in [58]. The definitions of G-reduction and G-expansion along with their properties are stated in [95].

More information on the Heisenberg group, Lie groups and so forth can be found in, e.g., [49] and the references cited there.

Chapter 2

Expanding endomorphisms and substitutions

In this chapter, we introduce the notion of a substitution on the space of sequences $\Sigma(\Gamma, \overline{\mathcal{A}})$. In order to do so, we have to consider expanding endomorphisms of the group Γ. This is done in the first section. We discuss the question which finitely generated groups possibly admit an expanding endomorphism $H : \Gamma \to \Gamma$.

In the second part, we give the definition of a (V, H)-substitution $S : \Sigma(\Gamma, \overline{\mathcal{A}}) \to \Sigma(\Gamma, \overline{\mathcal{A}})$, where $H : \Gamma \to \Gamma$ is an expanding endomorphism such that the index of the subgroup $H(\Gamma)$ is finite and V is a residue set of H. As a first step we shall investigate the dynamic properties of a substitution. Then we introduce the notion of a sequence which is generated by a substitution. As it will turn out later, these sequences are precisely the automatic sequences we are interested in.

Furthermore, we introduce a special kind of reduction maps, namely the so-called decimation. Along with the decimation of a sequence comes the kernel of a sequence $\underline{f} \in \Sigma(\Gamma, \mathcal{A})$. Loosely speaking the kernel of a sequence $\underline{f}$ consists of all sequences which can be obtained by repeated applications of the decimations. We shall show that a sequence is generated by a substitution if and only if the kernel of the sequence is finite.

Furthermore, we present two types of finite directed graphs. The first one is the so called graph of a substitution, and the second one is the kernel graph of a sequence $\underline{f}$. We discuss the relations between these two graphs.

2.1   Expanding endomorphisms

In this section we introduce the concept of an expanding group endomorphism of the finitely generated group Γ. We begin with the definition of expanding endomorphisms, followed by several examples of expanding endomorphisms.

The question on the existence of expanding endomorphisms of a group is a very delicate one. Its answer depends on the structure of the group as well as on the norm.

Finally, we investigate for which groups Γ there exists an expanding endomorphism $H : \Gamma \to \Gamma$ such that $H(\Gamma)$ is a subgroup of finite index.

From now on, $(\Gamma, \| \ \|)$ denotes a finitely generated group with a norm.

Definition 2.1.1. Let $H : (\Gamma, \| \ \|) \to (\Gamma, \| \ \|)$ be a group endomorphism. If there exists a $C > 1$, $C \in \mathbb{R}$, such that

$$\|H(\gamma)\| \geq C\|\gamma\|$$

holds for all $\gamma \in \Gamma$ then H is called *expanding* (w.r.t. the norm on Γ).

If we want to emphasize that an endomorphism H of Γ is expanding w.r.t. the norm induced by a generating set E of Γ, we say that H is expanding w.r.t. E.

Remarks.
1. The largest C satisfying the condition of Definition 2.1.1 is called the expansion ratio of H.

2. If $D_{\| \ \|}$ denotes the left invariant metric associated with the norm, then for an expanding endomorphism H we have $D(H(\gamma), H(\rho)) \geq CD(\gamma, \rho)$ for all $\gamma, \rho \in \Gamma$.

3. If H is expanding, then the n-th iterate, $H^{\circ n}$, of H is expanding, too. The expansion ratio of $H^{\circ n}$ is greater than or equal to C^n, where C is the expansion ratio of H.

4. If H is expanding, then H is injective.

5. If H is expanding (w.r.t. $\| \ \|$) with expansion ratio $C > 1$, then

$$\|H^{-1}(\gamma)\| \leq \frac{\|\gamma\|}{C}$$

holds for all $\gamma \in H(\Gamma) = \{H(\tau) \mid \tau \in \Gamma\}$.

6. The property of an endomorphism $H : \Gamma \to \Gamma$ to be expanding depends on the norm. E.g.: Let $\mathbb{Z}$ be the group of integers and let $H : \mathbb{Z} \to \mathbb{Z}$ be defined by $z \mapsto 2z$. Then H is expanding w.r.t. $E = \{1\}$ and the expansion ratio is equal to 2. If we consider $\mathbb{Z}$ with generating set $E' = \{2, 3\}$, then H is not expanding w.r.t. E'. We have $\|1\|_{E'} = 2$ and $\|H(1)\| = \|2\| = 1$.

7. If $H : \Gamma \to \Gamma$ is an endomorphism and E is a generating set such that $H(\Gamma) \cap E \neq \emptyset$, then H is not expanding w.r.t. E.

Examples.
1. Let $\Gamma = \langle x \rangle$, i.e., Γ is isomorphic to $\mathbb{Z}$. Then all expanding endomorphisms (w.r.t. $\{x\}$) are of the form $H(x^j) = x^{jd}$, where $d \in \mathbb{Z}$ and $|d| \geq 2$.

2. Let $\Gamma = \langle x, y \rangle$ be the free Abelian group, i.e., Γ is isomorphic to $\mathbb{Z}^2$. Any endomorphism $H : \langle x, y \rangle \to \langle x, y \rangle$ is given by a matrix $A_H \in \mathbb{Z}^{2 \times 2}$, and vice versa. Since any norm on the vector space $\mathbb{R}^2$ induces a norm on the group $\langle x, y \rangle$ (regarded as a subset of $\mathbb{R}^2$) the expanding property depends on the norm on $\mathbb{R}^2$. By a result in [114], there exists a norm on $\mathbb{R}^2$ such that H is expanding if all eigenvalues of the matrix A_H have modulus greater than 1.

3. Let L be the Heisenberg group introduced in Section 1.2 equipped with the left invariant Riemannian metric also introduced in 1.2. The map

$$H_{a,b} : L \to L$$

$$\begin{pmatrix} 1 & x & z \\ 0 & 1 & y \\ 0 & 0 & 1 \end{pmatrix} \mapsto \begin{pmatrix} 1 & ax & abz \\ 0 & 1 & by \\ 0 & 0 & 1 \end{pmatrix},$$

where $a, b \in \mathbb{R}$, is a group endomorphism of the Heisenberg group. If one chooses a, b such that $|a|, |b| > 1$, then $H_{a,b}$ is expanding w.r.t. the Riemannian metric. For $a, b \in \mathbb{Z}$, $H_{a,b}$ maps the discrete subgroup $L_\mathbb{Z}$ into $L_\mathbb{Z}$, and therefore we have for $a, b \in \mathbb{Z}$ and $|a|, |b| > 1$ that the map $H_{a,b} : (L_\mathbb{Z}, \| \ \|_r) \to (L_\mathbb{Z}, \| \ \|_r)$ is an expanding group endomorphism (w.r.t. the norm $\| \ \|_r$)

4. Let Γ be the free group generated by a, b equipped with the norm induced by the generating set $\{a, b\}$. Then the endomorphism $H : \Gamma \to \Gamma$ defined by the unique extension of the map $a \mapsto aa$ and $b \mapsto bb$ is an expanding endomorphism. Its expansion ratio is two.
   The unique extension of the map $a \mapsto ab^{-1}$ and $b \mapsto bb$ to an endomorphism H of the free group does not define an expanding map. We have $\|H(ab)\| = \|ab^{-1}bb\| = \|ab\|$ which shows that H is not an expanding map. Indeed, there exists no norm on Γ such that H is expanding.

As a first step towards a characterization of groups which admit expanding endomorphisms we begin with some simple observations.

Lemma 2.1.2. *Let Γ be a non-trivial finite group equipped with any norm. Then there exists no expanding endomorphism of Γ.*

Proof. Assume that $H : \Gamma \to \Gamma$ is an expanding (w.r.t any norm $\| \ \|$) endomorphism with expansion ratio $C > 1$. Since Γ is a finite group and H is injective, it follows that H is an automorphism of Γ. Since the automorphism group of Γ is finite there exists an $n_0 \in \mathbb{N}$ such that

$$H^{\circ n_0}(\gamma) = \gamma$$

holds for all $n \in \mathbb{N}$. We therefore conclude that

$$\|H^{\circ n_0}(\gamma)\| = \|\gamma\| \geq C^{n_0}\|\gamma\|$$

holds for all $\gamma \in \Gamma$; this is a contradiction. $\square$

Let $\mathrm{Tor}(\Gamma)$ denote the set of all elements of Γ which are of finite order. If $\mathrm{Tor}(\Gamma)$ is a subgroup of Γ and H is an expanding endomorphism, then, as a consequence of Lemma 2.1.2, either the cardinality of $\mathrm{Tor}(\Gamma)$ is infinite or $\mathrm{Tor}(\Gamma)$ is trivial, i.e., $\mathrm{Tor}(\Gamma) = \{e\}$. In case of a finitely generated Abelian group Γ we state the following result.

Lemma 2.1.3. *Let $\Gamma \neq \{e\}$ be a finitely generated Abelian group and $H : (\Gamma, \| \ \|) \to (\Gamma, \| \ \|)$ be an expanding endomorphism. Then Γ (as a group) is isomorphic to the group $\mathbb{Z}^n$ for some $n \in \mathbb{N}$.*

Proof. The set $\mathrm{Tor}(\Gamma)$ of Γ is a finite subgroup since Γ is finitely generated and Abelian. Moreover, $H(\mathrm{Tor}(\Gamma)) \subset \mathrm{Tor}(\Gamma)$ and by Lemma 2.1.2 it follows that $\mathrm{Tor}(\Gamma) = \{e\}$. By the classification theorem of finitely generated Abelian groups, we obtain that Γ is isomorphic to some $\mathbb{Z}^n$. $\square$

If Γ is finitely generated nilpotent, then the torsion subgroup is finite, see e.g. [123]. Thus if a finitely generated nilpotent group admits an expanding endomorphism the torsion subgroup must be trivial.

If one compares the examples of expanding endomorphisms given above, then one observes a striking difference between 1., 2., 3., and 4.. For the cases 1., 2., and 3. the subgroup $H(\Gamma)$ is of finite index, while for the case 4. the subgroup $H(\Gamma)$ is of infinite index, where the index of $H(\Gamma)$ is equal to the number of different equivalence classes given by the following equivalence relation: $\gamma, \gamma' \in \Gamma$ are called H-equivalent if $\gamma^{-1}\gamma' \in H(\Gamma)$.

For our study of substitutions we are only interested in expanding endomorphisms $H : \Gamma \to \Gamma$ such that the subgroup $H(\Gamma)$ is of finite index.

The next lemma shows that the existence of an expanding endomorphism of Γ implies that H-equivalent points cannot be too close together.

Lemma 2.1.4. *Let $H : (\Gamma, \| \ \|_E) \to (\Gamma, \| \ \|_E)$ be expanding (w.r.t. $E = \{\gamma_1, \ldots, \gamma_k\}$) with expansion ratio $C > 1$. Then the open ball $\overset{o}{B}_{\frac{C}{2}}(e)$ contains no H-equivalent points.*

Proof. Suppose there are $\gamma, \gamma' \in \overset{o}{B}_{\frac{C}{2}}(e)$ with $\gamma \neq \gamma'$ and $\gamma^{-1}\gamma' \in H(\Gamma)$, i.e., there exists a $\tau \in \Gamma$ with $\gamma^{-1}\gamma' = H(\tau)$. Then we estimate

$$C > \|\gamma\|_E + \|\gamma'\|_E \geq \|\gamma^{-1}\gamma'\|_E = \|H(\tau)\|_E \geq C\|\tau\|_E.$$

Since the norm only takes values in $\mathbb{N}$, we conclude that $\tau = e$ and therefore $\gamma = \gamma'$ which is a contradiction. $\square$

Let $\Gamma = \langle \gamma_1, \ldots, \gamma_k \rangle$ be a finitely generated group. The open ball of radius $r > 0$ with center τ is the set

$$\overset{o}{B}_r(\tau) = \{\gamma \mid \|\tau^{-1}\gamma\|_E < r\}.$$

Each ball of radius r centered at $\tau \in \Gamma$ contains only finitely many elements. Moreover, balls with different center but equal radius have equal cardinality.

Therefore it is sufficient to count the number of elements in the ball $B_r(e)$. In order to study groups which admit an expanding endomorphism we introduce the concept of polynomial growth of a group.

Definition 2.1.5. Let the group Γ be generated by the set $E = \{\gamma_1, \ldots, \gamma_k\}$. The group Γ has *polynomial growth* if there exist real numbers $K > 0$ and $D > 0$ such that

$$|B_r(e)| \leq Kr^D$$

holds for all $r \geq 0$.

Due two our preliminary remarks, see p. 9, the definition of polynomial growth does not depend on the choice of the generating set.

Examples.

1. Let $\Gamma = \langle x \rangle$, i.e., Γ is isomorphic to $\mathbb{Z}$. Then Γ has polynomial growth, since

$$|B_r(e)| = 2k + 1 \quad \text{for } r \in [k, k+1[\text{ and } k \in \mathbb{N}_0.$$

2. Let $\Gamma = \langle x, y \rangle$ be the free Abelian group, i.e., Γ is isomorphic to $\mathbb{Z}^2$, with generating set $E = \{x, y\}$. Then Γ has polynomial growth, since

$$|B_r(e)| = 2k^2 + 2k + 1 \quad \text{for } r \in [k, k+1[\text{ and } k \in \mathbb{N}_0.$$

3. Any finitely generated Abelian group has polynomial growth.

4. Let Γ be the free group generated by a, b equipped with the norm induced by the generating set $\{a, b\}$. Then Γ is not of polynomial growth, since

$$|B_r(e)| = 2 \cdot 3^k - 1 \quad \text{for } r \in [k, k+1[\text{ and } \in \mathbb{N}_0.$$

Moreover, any free group with more than two generators does not have polynomial growth. It rather has exponential growth, i.e.,

$$|B_r(e)| \geq K^r$$

for all $r > 1$ and some real constant $K > 1$.

Theorem 2.1.6. *Let Γ be a finitely generated group and $H : (\Gamma, \| \, \|_E) \to (\Gamma, \| \, \|_E)$ be an expanding endomorphism (w.r.t. the generating set E) with expansion ratio $C > 1$. If the index of the subgroup $H(\Gamma)$ is finite, then Γ has polynomial growth.*

Proof. By Lemma 2.1.4, we have that $\overset{o}{B}_{\frac{C}{2}}(e)$ contains no equivalent points. The n-th iterate $H^{\circ n}$ of H is expanding and $\overset{o}{B}_{\frac{C^n}{2}}(e)$ contains no $H^{\circ n}$-equivalent points. If $d \geq 1$ is the index of $H(\Gamma)$, then the index of $H^{\circ n}(\Gamma)$ is d^n, and, as a consequence of Lemma 2.1.4, we obtain

$$\left| \overset{o}{B}_{\frac{C^n}{2}}(e) \right| \leq d^n$$

for all $n \in \mathbb{N}$. For $\alpha = \frac{\log d}{\log C}$ the above inequality becomes

$$\left| \overset{o}{B}_{\frac{C^n}{2}} (e) \right| \leq 2^\alpha \left(\frac{C^n}{2} \right)^\alpha.$$

Since the function $r \mapsto \left| \overset{o}{B}_r(e) \right|$ is increasing, we have for all real numbers r such that $\frac{C^n}{2} \leq r < \frac{C^{n+1}}{2}$ the inequality

$$\left| \overset{o}{B}_r(e) \right| \leq \left| \overset{o}{B}_{\frac{C^{n+1}}{2}} (e) \right| \leq 2^\alpha \left(\frac{C^{n+1}}{2} \right)^\alpha = 2^\alpha \left(\frac{C^n}{2} \right)^\alpha C^\alpha \leq 2^\alpha dr^\alpha.$$

We therefore conclude that $\left| \overset{o}{B}_r(e) \right| \leq 2^\alpha dr^\alpha$ holds for all $r \geq 0$ which proves that Γ has the polynomial growth property.   □

Remarks.

1. Note that the above proof also shows that the index of $H(\Gamma)$ is not equal to 1, unless Γ is the trivial group $\{e\}$.

2. If Γ is the free group with more than two generators, then there exists no expanding endomorphism H such that $H(\Gamma)$ is of finite index.

In order to generalize the above result to arbitrary norms on Γ, we introduce the notion of a discrete norm.

Definition 2.1.7. Let Γ be a group and $\| \ \|$ a norm on Γ. The norm is called *discrete* if the set $\{ \|\gamma\| \mid \gamma \in \Gamma \}$ is a discrete subset of $\mathbb{R}_0^+$.

Remarks.

1. All previous examples of norms on groups are discrete norms.

2. Let $\Gamma = \langle x \rangle$, then

$$\|x^k\| = \frac{|k|}{1 + |k|}$$

   is a non-discrete norm on Γ.

3. If $\| \ \|$ is a bounded norm, i.e., $\|\Gamma\|$ is a bounded subset of $\mathbb{R}_0^+$, then there do not exist expanding endomorphisms (w.r.t. $\| \ \|$) on Γ.

Lemma 2.1.8. *Let Γ be a finitely generated group and let $H : (\Gamma, \| \ \|) \rightarrow (\Gamma, \| \ \|)$ be an expanding group endomorphism with expansion ratio $C > 1$. If the index of the subgroup $H(\Gamma)$ is finite and if the norm is discrete, then Γ has the polynomial growth property.*

Proof. Since $\| \ \|$ is a discrete norm, there exists an $r_0 > 0$ such that $|B_r(e)| = 1$ for all $0 \leq r < r_0$. Let d, $d \geq 1$, be the index of the subgroup $H(\Gamma) \subset \Gamma$. Then Lemma 2.1.4 generalizes to: The open ball $\overset{o}{B}_{r_0 \frac{C}{2}}(e)$ contains no equivalent points. With analogous arguments as in the proof of Theorem 2.1.5 one obtains that

$$\left| \overset{o}{B}_r(e) \right| \leq \left(\frac{2}{r_o} \right)^\alpha dr^\alpha, \tag{2.1}$$

where d is the index of $H(\Gamma)$ and $\alpha = \frac{\log d}{\log C}$, holds for all $r \geq 0$. Thus Γ has polynomial growth w.r.t. the norm $\| \ \|$. It remains to show that Γ has polynomial growth w.r.t. a generating set E. Let $E = \{\gamma_1, \ldots, \gamma_k\}$ be a generating set of Γ and let $\| \ \|_E$ be the induced norm. Let $\gamma = \gamma_{j_1}^{p_1} \ldots \gamma_{j_m}^{p_m}$ be a minimal word length representation of γ (w.r.t. E), then

$$\|\gamma\| = \|\gamma_{j_1}^{p_1} \ldots \gamma_{j_m}^{p_m}\| \leq \sum_{i=1}^{m} |p_i| \, \|\gamma_{j_i}\|.$$

Since E is a finite set, there exists a $K > 0$ such that $\|\gamma_j\| \leq K$ holds for all $\gamma_j \in E$. This gives

$$\|\gamma\| \leq K \sum_{i=0}^{m} |p_i| = K \|\gamma\|_E.$$

Therefore a $\| \ \|_E$-ball of radius r and center e is contained in the $\| \ \|$-ball of radius Kr and center e. By Equation (2.1), it follows that the cardinality of a $\| \ \|$-ball of radius r with center e is at most $\frac{2^\alpha}{r_0^\alpha} d(rK)^\alpha$, i.e., Γ has polynomial growth w.r.t. E. $\quad\square$

Due to a result of Gromov [88], we obtain that under the assumptions of Corollary 2.1.8 Γ contains a nilpotent subgroup of finite index.

After this excursion we focus our attention now on the substitutions.

2.2 Substitutions

We start with our example from the Introduction. Let $(t_n)_{n \in \mathbb{N}}$ be the Thue–Morse sequence, i.e.,

$$t_0 = 0$$
$$t_{2n} = t_n$$
$$t_{2n+1} = 1 - t_n.$$

Due to our observations, the Thue–Morse sequence is generated by a 'substitution' defined by $0 \mapsto 01$ and $1 \mapsto 10$. Starting with 0 we get a sequence of words

$$0 \mapsto 01 \mapsto 0110 \mapsto 01101001 \mapsto \cdots$$

which 'converges' to the Thue–Morse sequence. The goal of this section is to put this observation on a firm mathematical ground.

Let $\mathcal{A} = \{0, 1\}$ be a finite set and consider the sequence space $\Sigma(\mathbb{N}, \mathcal{A})$., i.e., the set of sequences over $\mathbb{N}$ with values in $\mathcal{A} = \{0, 1\}$. Although $\mathbb{N}$ with addition is not a group, we can consider it as a subset of the group $\mathbb{Z}$ with addition. In order to write addition of the group $\mathbb{Z}$ in a multiplicative way, we denote the elements of the group $\mathbb{Z}$ by x^j with $j \in \mathbb{Z}$. There is a natural projection of $p : \Sigma(\mathbb{Z}, \mathcal{A}) \to \Sigma(\mathbb{N}, \mathcal{A})$ defined by $p(\underline{f}) = \underline{f}_{|\mathbb{N}}$. There does not exist a natural embedding $\Sigma(\mathbb{N}, \mathcal{A})$ into $\Sigma(\mathbb{Z}, \mathcal{A})$; but there exists a natural embedding j of $\Sigma(\mathbb{N}, \mathcal{A})$ into $\Sigma(\mathbb{Z}, \overline{\mathcal{A}})$ defined by

$$i(\underline{f})(x^j) = \begin{cases} \underline{f}(x^j) & \text{if } j \geq 0 \\ \varnothing & \text{if } j < 0 \end{cases}$$

The map $H : \mathbb{Z} \to \mathbb{Z}$ defined by $H(x^j) = x^{2j}$ is a monomorphism of $\mathbb{Z}$ and $H(\mathbb{Z}) \subset \mathbb{Z}$ is a subgroup of index 2. Since $\mathbb{Z}$ is generated by x, there exists an induced norm on $\mathbb{Z}$ which we denote by $\| \ \|$ and which is defined by $\|x^j\| = |j|$. For x^j in $\mathbb{Z}$ we thus obtain $\|H(x^j)\| = 2\|x^j\|$, i.e., the monomorphism H is expanding w.r.t. the generating set $\{x\}$. Moreover, the set $\mathbb{N}$ is H-invariant, i.e., $H(\mathbb{N}) \subset \mathbb{N}$.

Since $H(\mathbb{Z})$ is a subgroup of index two, every element $x^j \in \mathbb{Z}$ can be uniquely written as $x^j = x^\epsilon H(x^{j'})$, where $\epsilon \in \{0, 1\}$. The set $V = \{x^0, x^1\}$ is called a residue set of H. Due to the choice of the residue set we even have that any element $x^j \in \mathbb{N}$ has a unique representation $x^j = x^\epsilon H(x^{j'})$ with $x^{j'} \in \mathbb{N}$.

As a next step we introduce the notion of a substitution. To this end, we consider the finite set $\mathcal{A} = \{0, 1\}$ and define a map $s : V \times \mathcal{A} \to \mathcal{A}$ by the following table:

	x^0	x^1
0	0	1
1	1	0

For the sake of simplicity, we write $s_{x^\epsilon} : \mathcal{A} \to \mathcal{A}$ for the map $a \mapsto s(x^\epsilon, a)$. The extensions to $\overline{\mathcal{A}}$ of the maps s_{x^ϵ} are defined by $s_{x^\epsilon}(\varnothing) = \varnothing$ and also denoted by s_{x^ϵ}.

We now consider the map $S : \Sigma(\mathbb{Z}, \overline{\mathcal{A}}) \to \Sigma(\mathbb{Z}, \overline{\mathcal{A}})$ defined by

$$S(\underline{f}) = \hat{s}_{x^0}(H_*(\underline{f})) \oplus \hat{s}_{x^1}((T_{x^1} \circ H)_*(\underline{f})).$$

Note that $\Sigma(\mathbb{Z}, \mathcal{A}) \subset \Sigma(\mathbb{Z}, \overline{\mathcal{A}})$ is S-invariant. Thus we can study S on $\Sigma(\mathbb{Z}, \mathcal{A})$. Moreover, due to our choice of V, we have that the set $\Sigma(\mathbb{N}, \mathcal{A})$ is S-invariant, too.

The map S is called a substitution. To justify the name substitution we study the orbit of the sequence $\underline{f}^0_{x^0}$ under the iteration of S. Since $\underline{f}^0_{x^0}$ denotes the sequence with value 0 at x^0 and $\varnothing$ otherwise, we obtain the following orbit:

$$\underline{f}^0_{x^0} = (\ldots \varnothing \varnothing \varnothing . 0 \varnothing \varnothing \varnothing \varnothing \varnothing \ldots)$$
$$S(\underline{f}^0_{x^0}) = (\ldots \varnothing \varnothing \varnothing . 0 1 \varnothing \varnothing \varnothing \varnothing \ldots)$$
$$S^{\circ 2}(\underline{f}^0_{x^0}) = (\ldots \varnothing \varnothing \varnothing . 0 1 1 0 \varnothing \varnothing \ldots),$$

where "." indicates the lower left corner of the zero position. The map $S_{|\mathbb{N}} = p \circ S \circ i$ is a substitution on $\Sigma(\mathbb{N}, \mathcal{A})$ which exactly models the generation of the Thue–Morse sequence. Moreover, the Thue–Morse sequence, $\underline{t}$, is a fixed point of the substitution $S_{|\mathbb{N}} = p \circ S \circ i$, i.e., we have $S_{|\mathbb{N}}(\underline{t}) = \underline{t}$. If we set $u_n = 1 - t_n$ for all $n \in \mathbb{N}$ then the sequence $\underline{u} = (u_n)$ is another fixed point of $S_{|\mathbb{N}} = p \circ S \circ i$. These are the only fixed points of $S_{|\mathbb{N}} = p \circ S \circ i$.

The substitution S on $\Sigma(\mathbb{Z}, \overline{\mathcal{A}})$ has three fixed points; the empty sequence, the embedded Thue–Morse sequence $i(\underline{t})$ and the embedded sequence

$$\underline{u} = \bigoplus_{j \geq 0} x^j \overline{t}_j,$$

where $\overline{0} = 1$ and $\overline{1} = 0$. Moreover, S on $\Sigma(\mathbb{Z}, \overline{\mathcal{A}})$ has two periodic points $\{\underline{f}_1, S(\underline{f}_1)\}$, $\{\underline{f}_2, S(\underline{f}_2)\}$ of period two, where

$$\underline{f}_1 = \bigoplus_{j < 0} x^j\, t_{-j-1} \oplus \bigoplus_{j \geq 0} x^j\, t_j, \quad \underline{f}_2 = \bigoplus_{j < 0} x^j\, t_{-j-1} \oplus \bigoplus_{j \geq 0} x^j\, u_j$$

The substitution S on $\Sigma(\mathbb{Z}, \mathcal{A})$ has two periodic orbits of period two and no fixed points.

After this introductory example we start with the general theory. As we have seen, it is the expanding endomorphism $H(x^j) = x^{2j}$ and the finite residue set V which allow us to define a substitution. We therefore deal with expanding endomorphisms and residue sets first.

Definition 2.2.1. Let $(\Gamma, \| \ \|)$ be a finitely generated group with a norm, and let H be an expanding endomorphism such that $H(\Gamma)$ is a subgroup of finite index. A set $V \subset \Gamma$ is called a *(left)residue set* (w.r.t. H) if the following holds.

1. The unit e of Γ is in V.

2. For any $\gamma \in \Gamma$ there exist a unique $v \in V$ and a unique $\gamma' \in \Gamma$ such that

$$\gamma = vH(\gamma').$$

For the sake of simplicity, we often call V a residue set (of H).

Remarks.

1. The condition $e \in V$ is not really necessary for the definition of a residue set. However, for all what follows it is very convenient to ensure that e belongs to a residue set.

2. The number of elements in a residue set is equal to the index of the subgroup $H(\Gamma)$. Therefore all residue sets have the same cardinality.

3. Let V be residue set (w.r.t. H), and define the sequence $(V_n)_{n \in \mathbb{N}}$ as $V_0 = V$ and for $n \geq 1$ set

$$V_n = VH(V_{n-1}) = \{vH(w) \mid v \in V, w \in V_{n-1}\}.$$

Then V_n is a residue set w.r.t. $H^{\circ n}$.

Examples.

1. Let $\Gamma = \langle x \rangle$, i.e., Γ is isomorphic to $\mathbb{Z}$. Then all expanding endomorphisms (w.r.t. the norm induced by $\{x\}$) are of the form $H(x^j) = x^{jd}$, where $d \in \mathbb{Z}$ and $|d| \geq 2$. A residue set (of $H(x) = x^{dj}$) is given by $V = \{0, 1, \ldots, |d| - 1\}$.

2. Let $\Gamma = \langle x, y \rangle$ be the free Abelian group, i.e., Γ is isomorphic to the additive group $\mathbb{Z}^2$. Any endomorphism $H : \langle x, y \rangle \to \langle x, y \rangle$ is given by a matrix $A \in \mathbb{Z}^{2 \times 2}$, and vice versa. Since any norm on the vector space $\mathbb{R}^2$ induces a norm on the group $\langle x, y \rangle$ (regarded as a subset of $\mathbb{R}^2$) the expanding property depends on the norm chosen on $\mathbb{R}^2$. The endomorphism $H : \langle x, y \rangle \to \langle x, y \rangle$ defined by the matrix

$$\begin{pmatrix} 1 & -1 \\ 1 & 1 \end{pmatrix}$$

is expanding w.r.t. the euclidean metric; it is not expanding w.r.t. the metric induced by the generating set $\{x, y\}$. A residue set is given by $V = \{x^0 y^0, x^1 y^0\}$.

3. Let Γ_3 be the discrete Heisenberg group and consider the endomorphism $H_{2,2}$, see Example 3, p. 21,

$$H_{2,2} : L \to L$$

$$\begin{pmatrix} 1 & x & z \\ 0 & 1 & y \\ 0 & 0 & 1 \end{pmatrix} \mapsto \begin{pmatrix} 1 & 2x & 4z \\ 0 & 1 & 2y \\ 0 & 0 & 1 \end{pmatrix}.$$

Then H is an expanding endomorphism (w.r.t. the norm $\| \ \|_r$). A (left)-residue set is given by

$$V = \{b^{\epsilon_1} a^{\epsilon_2} c^{\epsilon_3} \mid \epsilon_1, \epsilon_2 \in \{0, 1\}, \epsilon_3 \in \{0, 1, 2, 3\}\}.$$

By Lemma 1.2.1 each $\gamma \in \Gamma_3$ has a unique representation as $\gamma = a^n b^m c^p$ and therefore $H(\gamma) = b^{2n} a^{2m} c^{4p}$. If we write $n = 2n' + \epsilon_1$, $m = 2m' + \epsilon_2$, where $\epsilon_1, \epsilon_2 \in \{0, 1\}$ and $n', m' \in \mathbb{Z}$, then we have

$$\gamma = a^n b^m c^p = a^{2n'+\epsilon_1} b^{2m'+\epsilon_2} c^p = b^{\epsilon_1} a^{\epsilon_2} a^{2n'} b^{2m'} c^{p+2n'\epsilon_2},$$

where the last equation is due to Lemma 1.2.1. Since $p - 2n'\epsilon_2 = 4p' + \epsilon_3$, where $\epsilon_3 \in \{0, 1, 2, 3\}$ and $p' \in \mathbb{Z}$, we obtain

$$\gamma = b^{\epsilon_1} a^{\epsilon_2} c^{\epsilon_3} a^{2n'} b^{2m'} c^{4p'} = b^{\epsilon_1} a^{\epsilon_2} c^{\epsilon_3} H(b^{n'} a^{m'} c^{p'}).$$

The uniqueness property of the representation of γ as $\gamma = a^n b^m c^p$ yields that V is a residue set.

Definition 2.2.2. Let $(\Gamma, \| \ \|)$ be a finitely generated group with a norm, and let H be an expanding endomorphism such that $H(\Gamma)$ is a subgroup of finite index. Moreover, let V be a residue set (w.r.t. H). The maps $\kappa = \kappa_{H,V} : \Gamma \to \Gamma$ and $\zeta = \zeta_{H,V} : \Gamma \to V$ defined by

$$\gamma = \zeta(\gamma) H(\kappa(\gamma))$$

are called *image-part-map* and *remainder-map*, respectively.

The image-part-map and the remainder-map define a kind of euclidian algorithm on Γ. If Γ is equipped with a discrete norm, see Definition 2.1.7, then the dynamics of $\kappa_{H,V}$ is described by the next lemma.

Lemma 2.2.3. *Let $(\Gamma, \| \ \|)$ be a finitely generated group with a discrete norm, and let H be an expanding endomorphism such that $H(\Gamma)$ is a subgroup of finite index. For every residue set V there exists an $R_V > 0$ such that $\kappa_{H,V}$ maps the closed $\| \ \|$-ball $B_{R_V}(e) = \{\gamma \in \Gamma \mid \|\gamma\| \leq R_V\}$ into itself.*
For each $\gamma \in \Gamma$, there exists an $n \in \mathbb{N}$ such that $\kappa_{H,V}^{\circ n}(\gamma) \in B_{R_V}(e)$.

Proof. We have $\zeta_{H,V}(\gamma)^{-1}\gamma = H(\kappa_{H,V}(\gamma))$, and therefore

$$\|\kappa_{H,V}(\gamma)\| \leq \frac{r_V + \|\gamma\|}{C},$$

where $r_V = \max\{\|v\| \mid v \in V\}$. For $\|\gamma\| \leq R_V = \frac{r_V}{C-1}$ we have

$$\|\kappa(\gamma)\| \leq R_V.$$

If $\|\gamma\| > R_V$, then $\|\kappa(\gamma)\| < \|\gamma\|$. Since the norm is discrete, there exists an $n \in \mathbb{N}$ such that $\|\kappa_{H,V}^{\circ n}(\gamma)\| \leq R_V$. This proves the second assertion. $\square$

By our previous results, see Section 2.1, we see that under the conditions of Lemma 2.2.3, $|B_{R_V}(e)|$ is finite. Since κ maps the ball $B_{R_V}(e)$ into itself, κ has only finitely many periodic points all of which are contained in $B_{R_V}(e)$. The set of periodic points of κ is denoted by $\operatorname{Per} \kappa$, i.e., $\operatorname{Per} \kappa = \{\gamma \mid \text{there exists an } n \in \mathbb{N} \text{ such that } \kappa^{\circ n}(\gamma) = \gamma\}$. As usual, the smallest positive n with $\kappa^{\circ n}(\gamma) = \gamma$ is called the period of $\gamma \in \operatorname{Per} \kappa$. If the period of $\gamma \in \operatorname{Per} \kappa$ equals 1, then we call γ a fixed point of κ. Obviously, we have $e \in \operatorname{Per} \kappa$, since $\kappa(e) = e$.

Given any $\gamma \in \Gamma$ there exists an n such that $\kappa^{\circ n}(\gamma)$ is a periodic point of κ, i.e., there exists an $m \in \mathbb{N}$ such that $\kappa^{\circ(n+m)}(\gamma) = \kappa^{\circ n}(\gamma)$. In other words, every γ is a preperiodic point of κ.

As a consequence of Lemma 2.2.3 we have the following corollary.

Corollary 2.2.4. *Let* $(\Gamma, \| \; \|)$ *be a finitely generated group with a discrete norm, and let H be an expanding endomorphism such that $H(\Gamma)$ is a subgroup of finite index. Furthermore, let V be a residue set. Then for all $R > 0$ there exists an $n \in \mathbb{N}$ such that $\kappa_{H,V}^{\circ -n}(\text{Per}\,\kappa_{H,V}) \supset B_R(e)$.*

Among the set of residue sets are special residue sets, the so-called complete digit sets.

Definition 2.2.5. Let $(\Gamma, \| \; \|)$ be a finitely generated group and let H be an expanding endomorphism such that $H(\Gamma)$ is a subgroup of finite index. A residue set V (w.r.t. H) is called a *complete digit set (of H)* if $\text{Per}\,\kappa_{H,V} = \{e\}$.

If V is a complete digit set for H and Γ, then the euclidian algorithm defined by $\kappa_{H,V}$ and $\zeta_{H,V}$ terminates at e for all $\gamma \in \Gamma$.

Corollary 2.2.6. *Let* $(\Gamma, \| \; \|)$ *be a finitely generated group, and let H be an expanding endomorphism such that $H(\Gamma)$ is a subgroup of finite index. A residue set V is a complete digit set if and only if each $\gamma \in \Gamma \setminus \{e\}$ has a finite representation as*

$$\gamma = v_{i_0} H(v_{i_1}) \ldots H^{\circ n}(v_{i_n})$$

with $v_{i_j} \in V$ for $j = 0, \ldots, n$ and $v_{i_n} \neq e$.

Proof. Let V be a residue set and let ζ and κ be the associated remainder-map and image-part-map, respectively.

Suppose that V is a complete digit set. For every $\gamma \in \Gamma \setminus \{e\}$ there exists an $n \in \mathbb{N}$ such that $\kappa^{\circ n}(\gamma) = e$. Therefore there exist finite sequences $(\gamma_k = \kappa^{\circ k}(\gamma))_{k=0,\ldots,n-1}$ and $(v_k = \zeta(\gamma_k))_{k=0,\ldots,n-2}$ such that $\gamma_k = \zeta(\gamma_k)H(\kappa(\gamma_k))$. We therefore have

$$\gamma = \zeta(\gamma_0)H(\zeta(\gamma_1)) \ldots H^{\circ n-1}(\zeta(\gamma_{k-1})),$$

which proves the first assertion.

The second assertion follows from

$$\kappa(\gamma) = \kappa(v_{i_0}H(v_{i_1})H(v_{i_2}) \ldots H^{\circ n}(v_{i_n})) = v_{i_1} \ldots H^{\circ n-1}(v_{i_n}). \qquad \square$$

Examples.

1. Let $\Gamma = \langle x \rangle$, i.e., Γ is isomorphic to $\mathbb{Z}$. Then the map $H(x^j) = x^{2j}$ is expanding (w.r.t. $E = \{x\}$). The set $V = \{x^0, x^1\}$ is a residue set of H but not a complete digit set. We have $\text{Per}\,\kappa_{H,V} = \{x^{-1}, x^0\}$, and all elements of $\text{Per}\,\kappa_{H,V}$ are fixed points.

   Moreover, there exists no residue set V such that V is a complete digit set for $H(x^j) = x^{2j}$.

2. Let $\Gamma = \langle x \rangle$ and $H(x^j) = x^{-2j}$. Then H is expanding (w.r.t. $E = \{x\}$) and the set $V = \{x^0, x^1\}$ is a complete digit set.

3. Let $\Gamma = \langle x \rangle$, then $H(x^j) = x^{3j}$ is expanding. The set $V = \{x^0, x^1, x^2\}$ is a residue set which is not a complete digit set. The periodic points are x^{-1} and x^0 and both are fixed points.

   The set $V_1 = \{x^{-1}, x^0, x^{13}\}$ is a residue set for $H(x^j) = x^{3j}$. Then x^0 is a fixed point of the associated image-part-map κ, the point x^{-2} is a periodic point of period 3, i.e., $\kappa(x^{-2}) = x^{-5}$, $\kappa(x^{-5}) = x^{-6}$, and $\kappa(x^{-6}) = x^{-2}$.

   The set $V_2 = \{x^{-1}, x^0, x^1\}$ is a complete digit set for H.

4. For $\Gamma = \langle x, y \rangle$, the free Abelian group with two generators, the map $H(x^i y^j) = x^{2i} y^{2j}$ is expanding (w.r.t. $E = \{x, y\}$). A residue set V of H is given by $V = \{x^0 y^0, x^1 y^0, x^0 y^1, x^1 y^1\}$, however, V is not a complete residue set. Moreover, there does not exist a complete digit set for H.

   Let $H_b(x^i y^j) = x_1^{-ib-j} x_2^{-bj-i}$ is expanding (w.r.t, the norm induced by the euclidian metric on $\mathbb{R}^2$). The set $V = \{x_1^0, x_1^1, x_1^2, \ldots, x_1^{b^2}\}$ is a complete digit set, cf. [103].

5. Let $L_{\mathbb{Z}}$ be the discrete Heisenberg group introduced in Section 1.2. Consider

$$H : L \to L$$

$$\begin{pmatrix} 1 & x & z \\ 0 & 1 & y \\ 0 & 0 & 1 \end{pmatrix} \mapsto \begin{pmatrix} 1 & 2x & 4z \\ 0 & 1 & 2y \\ 0 & 0 & 1 \end{pmatrix}.$$

   Then H is an expanding endomorphism (w.r.t. the norm $\| \|_r$). A (left)-residue set is given by $V = \{b^{\epsilon_1} a^{\epsilon_2} c^{\epsilon_3} \mid \epsilon_1, \epsilon_2 \in \{0, 1\} \text{ and } \epsilon_3 \in \{0, 1, 2, 3\}\}$. The residue set V is not a complete digit set. The fixed points of κ are e, a^{-1}, b^{-1}, $c^{-1}, b^{-1}c^{-1}, a^{-1}c^{-1}, b^{-1}a^{-1}c^{-1}$. Moreover, these are the only periodic points of κ.

The next theorem provides a simple criterion for the existence of complete digit sets.

Theorem 2.2.7. *Let $H : \Gamma \to \Gamma$ be expanding w.r.t. the discrete norm $\| \|$ with expansion ratio $C > 2$ and let $H(\Gamma)$ be of index $d \in \mathbb{N}$. Then there exists a residue set that is a complete digit set (of H).*

Proof. Let $V = \{v_1 = e, v_2, \ldots, v_d\}$ be an arbitrary residue set (of H). For each $j = 1, 2, \ldots, d$ there exists a $\xi_j \in v_j H(\Gamma)$ such that

$$\|\xi_j\| \leq \|v_j H(\gamma)\|$$

holds for all $\gamma \in \Gamma$. The set $\Xi = \{\xi_1 = e, \xi_2, \ldots, \xi_d\}$ is a residue set and a complete digit set. Due to Definition 2.2.5 we have to show that $\operatorname{Per} \kappa_{\Xi, H} = \{e\}$. Suppose that $\gamma_0 \in \operatorname{Per} \kappa_{H, \Xi}$ is a non-trivial periodic point of period n. Then we have

$$\gamma_0 = \alpha_0 H(\alpha_1) H^{\circ 2}(\alpha_2) \ldots H^{\circ n-1}(\alpha_{n-1}) H^{\circ n}(\gamma_0),$$

where $\alpha_i \in \Xi, i = 0, \ldots, n-1$. Note that $\gamma_1 = \kappa_{\Xi, H}(\gamma_0)$ is also a periodic point of $\kappa_{\Xi,, H}$ and

$$\gamma_1 = \alpha_1 H(\alpha_2) \ldots H^{\circ n-2}(\alpha_{n-1}) H^{\circ n-1}(\alpha_0) H(\gamma_1).$$

This observation allows us to choose γ_0 in such a way that $\|\alpha_0\| \geq \|\alpha_j\|$ for all $j = 0, \ldots, n-1$. We also have

$$\gamma_0 = H^{-1}\big(\alpha_{n-1}^{-1} H^{-1}(\ldots H^{-1}(\alpha_1^{-1} H^{-1}(\alpha_0^{-1} \gamma_0))) \ldots)\big),$$

which yields the estimate

$$\|\gamma_0\| \leq \frac{\|\alpha_n\|}{C} + \frac{\|\alpha_{n-1}\|}{C^2} + \cdots + \frac{\|\alpha_0\| + \|\gamma_0\|}{C^n}.$$

Due to our assumption, we have that $\|\alpha_0\| \geq \|\alpha_j\|$ for all $j = 0, \ldots, n-1$. This leads to

$$\|\gamma\| \leq \frac{\|\alpha_0\|}{C-1}.$$

Since $C > 2$ we have that $\|\gamma\| < \|\alpha_0\|$ and $\gamma_0 = \alpha_0 H(\gamma')$. This contradicts the choice of the residue set Ξ. $\qquad\square$

Remarks.

1. If $H : \Gamma \to \Gamma$ is an expanding endomorphism with expansion ratio $C > 1$, then there exists $n_0 \in \mathbb{N}$ such that $H^{\circ n_0}$ has an expansion ratio of at least $C^{n_0} > 2$. Due to the above theorem, there exists a complete digit set of $H^{\circ n_0}$.

2. The estimate $C > 2$ is sharp in the sense that there exists an expanding group endomorphism H with expansion ratio $C = 2$ such that H has no complete digit set, see Example 1, p. 30.

   On the other hand, Example 2, p. 31, shows that there exist expanding maps H with expansion ratio $C = 2$ such that H has a complete digit set.

   Even for the case $C < 2$ there exists examples of expanding maps with a complete digit set. E.g., the endomorphism $H(x^i y^j) = x^{-1-j} y^{i-j}$ which is the multiplication by $-1 + \iota$, has expansion ratio $C = \sqrt{2}$ (w.r.t. euclidian metric) and the complete digit set $V = \{x^0 y^0, x^1 y^0\}$.

We are now prepared to state the definition of substitutions.

Definition 2.2.8. Let $(\Gamma, \| \, \|)$ be a finitely generated group and $H : \Gamma \to \Gamma$ an expanding endomorphism (w.r.t. $\| \, \|$) such that $H(\Gamma)$ is of finite index. Furthermore, let V be a residue set (of H) and $\mathcal{A}$ a finite set, and let $s : V \times \mathcal{A} \to \mathcal{A}$ be a map. Then s induces a map $S : \Sigma(\Gamma, \mathcal{A}) \to \Sigma(\Gamma, \mathcal{A})$ defined by

$$S(\underline{f})(\gamma) = s(v, \underline{f}(\rho)),$$

where $\gamma = vH(\rho)$ is the unique representation relative to H and V. The map S is called a (V, H)-*substitution*.

Note that a substitution has a canonical extension, also denoted by S, to $\Sigma(\Gamma, \overline{\mathcal{A}})$. This extension is defined by $s(v, \varnothing) = \varnothing$ for all $v \in V$. With this extension the set $\Sigma(\Gamma, \mathcal{A}) \subset \Sigma(\Gamma, \overline{\mathcal{A}})$ is an invariant set of S.

For $v \in V$, we denote the map $\overline{\mathcal{A}} \ni a \mapsto s(v, a) \in \overline{\mathcal{A}}$ by s_v. If the map H and the set V are clear from the context, we simply speak of the substitution S.

In terms of formal series, and it is here where we have to consider S as a map on $\Sigma(\Gamma, \overline{\mathcal{A}})$, we can write S in the following way:

$$S(\underline{f}) = S(\bigoplus f_\gamma \gamma) = \bigoplus_{v \in V} (T_v \circ H)_* \circ \widehat{s_v}(\underline{f}),$$

where $T_v : \Gamma \to \Gamma, \gamma \mapsto v\gamma$ are left-translations. If we use the short form for the map $(T_\tau)_*$, we can rewrite $S(\underline{f})$ as

$$S(\underline{f}) = \bigoplus_{v \in V} vH_*(\widehat{s_v}(\underline{f})).$$

More informally, we can write

$$S(\underline{f}) = \bigoplus_{v \in V} \bigoplus_{\gamma \in \Gamma} vH(\gamma)s(v, f_\gamma).$$

If the sum $\underline{f} \oplus \underline{g}$ is defined, then we have $S(\underline{f} \oplus \underline{g}) = S(\underline{f}) \oplus S(\underline{g})$.

The next lemma describes the dynamics of a substitution S.

Lemma 2.2.9. *Let* $S : \Sigma(\Gamma, \overline{\mathcal{A}}) \to \Sigma(\Gamma, \overline{\mathcal{A}})$ *be a* (V, H)-*substitution. Then the* ω-*limit set of* $\underline{f} \in \Sigma(\Gamma, \overline{\mathcal{A}})$ *under* S, *i.e., the set*

$$\omega(\underline{f}) = \{\underline{g} \mid \text{there exists } (n_j)_{j \in \mathbb{N}} \text{ with } \lim_{j \to \infty} n_j = \infty \text{ such that } \lim_{j \to \infty} S^{\circ n_j}(\underline{f}) = \underline{g}\},$$

is a periodic orbit of S.

Proof. Let κ be the image-part-map associated with H and V and let $\text{Per}\,\kappa$ be the set of periodic points of κ. Let $\underline{f} \in \Sigma(\Gamma, \overline{\mathcal{A}})$ and denote by $R(\underline{f})$ the restriction of $\underline{f}$ on $\text{Per}\,\kappa$, i.e.,

$$R(\underline{f}) = \bigoplus_{\gamma \in \text{Per}\,\kappa} \gamma f_\gamma,$$

with the usual convention $R(\underline{f})(\gamma) = \varnothing$ for $\gamma \notin \operatorname{Per} \kappa$.

Let Δ be a distance on $\Sigma(\Gamma, \overline{\mathcal{A}})$, see Equation (1.1), We begin by proving

$$\lim_{n \to \infty} \Delta(S^{\circ n}(\underline{f}), S^{\circ n}(R(\underline{f}))) = 0.$$

This means that the ω-limit set of $\underline{f}$ is determined by the ω-limit set of the restriction of $\underline{f}$ on $\operatorname{Per} \kappa$.

For the first iterate $S(R(\underline{f}))$ we obtain

$$S(R(\underline{f})) = \bigoplus_{v \in V} \bigoplus_{\gamma \in \operatorname{Per} \kappa} vH(\gamma) s(v, R(\underline{f})_\gamma) = \bigoplus_{\gamma \in \kappa^{-1}(\operatorname{Per} \kappa)} \gamma \, S(R(\underline{f}))(\gamma),$$

and for $S(\underline{f})$ we obtain

$$S(\underline{f}) = S(R(\underline{f})) \oplus S(\bigoplus_{\gamma \notin \operatorname{Per} \kappa} \gamma \, f_\gamma) = S(R(\underline{f})) \oplus \bigoplus_{\gamma \notin \kappa^{-1}(\operatorname{Per} \kappa)} \gamma \, S(\underline{f})(\gamma).$$

Therefore we have $S(R(\underline{f}))(\gamma) = S(\underline{f})(\gamma)$ for all $\gamma \in \kappa^{-1}(\operatorname{Per} \kappa)$. By Corollary 2.2.4, we easily obtain

$$S^{\circ n}(R(\underline{f}))(\gamma) = S^{\circ n}(\underline{f})(\gamma)$$

for all $\gamma \in \kappa^{\circ -n}(\operatorname{Per} \kappa)$. This proves the first step. In other words, the ω-limit sets of $\underline{f}$ and $R(\underline{f})$ coincide. Since the set of maps from $\operatorname{Per} \kappa$ to $\overline{\mathcal{A}}$ is finite, there exist $n_0, k \in \mathbb{N}$ such that

$$R(S^{\circ n_0 + k}(\underline{f})) = R(S^{\circ n_0}(\underline{f})),$$

which implies that the ω-limit set of $\underline{f}$ is a periodic orbit. $\square$

Remarks.

1. Any (V, H)-substitution on $\Sigma(\Gamma, \overline{\mathcal{A}})$ has the empty sequence as a fixed point.

2. If $S : \Sigma(\Gamma, \mathcal{A}) \to \Sigma(\Gamma, \mathcal{A})$ is a (V, H)-substitution, then the n-th iterate $S^{\circ n}$ of S is a $(V_n, H^{\circ n})$-substitution. Thus, by considering sufficiently high iterates of H and S we can always assume that $\operatorname{Per} \kappa$ consists of fixed points.

3. Let V be a residue set (of H) such that the associated image-part-map κ has only fixed points as periodic points. If S is a (V, H)-substitution, then the number of fixed points of S is given as the product

$$\prod_{\xi \in \operatorname{Per} \kappa} |\{a \mid s_v(a) = a, \text{ where } \xi = vH(\xi)\}| \, .$$

   If V is a complete digit set (of H), then the number of fixed points of a (V, H)-substitution is given by

$$|\{a \mid s_e(a) = a\}| \, .$$

Under certain conditions one can restrict a substitution S on a subset of $\Sigma(\Gamma, \mathcal{A})$.

Definition 2.2.10. Let $\Gamma' \subset \Gamma$ be a subset of Γ. Then Γ' is called (V, H)-*substitution-invariant* if

$$\Gamma' = \bigcup_{v \in V} vH(\Gamma'),$$

where $vH(\Gamma') = \{vH(\gamma') \mid \gamma' \in \Gamma'\}$.

Remarks.

1. The set Γ is a (V, H)-substitution-invariant set for all residue sets V. In fact, it is the only non-trivial set with this property.

2. Let Γ' be a (V, H)-substitution-invariant set and $e \in \Gamma'$, then the restriction $\kappa_{|\Gamma'} : \Gamma' \to \Gamma'$ is well defined. Following Definition 2.2.5, we say that V is a complete digit set of Γ' if and only if $\operatorname{Per} \kappa_{|\Gamma'} = \{e\}$.

3. Let $\kappa = \kappa_{H,V}$ denote the remainder-map, see Definition 2.2.2 and let ω be a subset of Γ, then

$$\Omega = \bigcup_{n \in \mathbb{N}} \kappa^{\circ -n}(\omega)$$

is the smallest (V, H)-substitution-invariant set containing ω.

4. If Γ' is (V, H)-substitution-invariant then the restriction $S_{|\Gamma'} : \Sigma(\Gamma', \mathcal{A}) \to \Sigma(\Gamma', \mathcal{A})$ of a (V, H)-substitution $S : \Sigma(\Gamma, \mathcal{A}) \to \Sigma(\Gamma, \mathcal{A})$ is well defined. If there is no risk of confusion, we simply speak of the substitution S on Γ'. Lemma 2.2.9 applies also for a substitution S on Γ'.

5. Let $\xi \in \operatorname{Per} \kappa_{H,V}$ be a fixed point of κ, then Γ_ξ denotes the smallest (V, H)-substitution-invariant subset containing ξ. By 2., one has $\Gamma_\xi = \bigcup_{n \in \mathbb{N}} \kappa^{\circ -n}(\xi)$.

   As a special case we note that the set Γ_e is (V, H)-substitution-invariant and V is a complete digit set of Γ_e.

6. If $\Gamma = \mathbb{Z}$ and $H(x^j) = x^{dj}$ with $d \geq 2$, then $V = \{x^0, x^1, \ldots, x^{d-1}\}$ is not a complete digit set for Γ. However, the subset $\mathbb{N} \subset \mathbb{Z}$ is (V, H)-substitution-invariant and V is a complete digit set for $\mathbb{N}$.

Definition 2.2.11. Let $\underline{f} \in \Sigma(\Gamma, \mathcal{A})$. If there exist a finite set $\mathcal{B}$, a map $\theta : \mathcal{B} \to \mathcal{A}$ and a (V, H)-substitution $S : \Sigma(\Gamma, \mathcal{B}) \to \Sigma(\Gamma, \mathcal{B})$ with fixed point $\underline{F} \in \Sigma(\Gamma, \mathcal{B})$ such that

$$\hat{\theta}(\underline{F}(\gamma)) = \underline{f}(\gamma)$$

holds for all $\gamma \in \Gamma$, then $\underline{f}$ is *generated by a (V, H)-substitution*.

Remarks.

1. The above definition immediately generalizes to sequences $\underline{f} \in \Sigma(\Gamma', \mathcal{A})$, where Γ' is a (V, H)-substitution-invariant set.

2. If $\underline{f} \in \Sigma(\Gamma, \mathcal{A})$ is generated by a substitution $S : \Sigma(\Gamma, \mathcal{A}) \to \Sigma(\Gamma, \mathcal{A})$, then $\underline{f}$ regarded as an element of $\Sigma(\Gamma, \overline{\mathcal{A}})$ is generated by the natural extension of S to $\Sigma(\Gamma, \overline{\mathcal{A}})$.

Examples.

1. Let $\Gamma = \langle x \rangle = \mathbb{Z}$ and $H(x^j) = x^{2j}$ with residue set $V = \{x^0, x^1\}$. Then $\Gamma_{x^0} = \{x^j \mid j \geq 0\} = \mathbb{N}$ is (V, H)-substitution-invariant and the Thue–Morse sequence, regarded as an element of $\Sigma(\mathbb{N}, \{0, 1\})$ is generated by the substitution S defined by

	x^0	x^1
0	0	1
1	1	0

The finite set $\overline{\mathcal{B}}$ from Definition 2.2.11 is equal to $\overline{\mathcal{A}} = \{0, 1, \varnothing\}$, the map θ is the identity. As already noted, the substitution S restricted on $\Sigma(\mathbb{N}, \{0, 1\})$ has two different fixed points. In the light of Lemma 2.2.9, these two fixed points are given as limits, i.e.

$$\underline{t} = \lim_{n \to \infty} S^{\circ n}(x^0 0),$$

$$\underline{u} = \lim_{n \to \infty} S^{\circ n}(x^0 1).$$

The substitution S considered as a map from $\Sigma(\Gamma, \mathcal{A})$ has no fixed point, it rather has four periodic points of period two.

2. Let $H(x^j) = x^{2j}$ and $V = \{x^0, x^1\}$, let $\mathcal{A} = \{0, 1, 2, 3\}$ and define a substitution by

	x^0	x^1
0	0	1
1	2	1
2	1	3
3	3	3

Since $\mathrm{Per}\, \kappa = \{-1, 0\}$ consists of fixed points only, the number of fixed points of S on $\Sigma(\mathbb{Z}, \mathcal{A})$ is given by

$$\left|\{a \mid s_{x^0}(a) = a\}\right| \left|\{a \mid s_{x^1}(a) = a\}\right| = |\{0, 3\}|\, |\{1, 3\}|$$

which is 4. The fixed points $\underline{f}_{(\alpha, \beta)}$, where $(\alpha, \beta) \in \{0, 3\} \times \{1, 3\}$ are given by

$$\underline{f}_{(\alpha, \beta)} = \lim_{n \to \infty} S^{\circ n}(x^{-1}\beta \oplus x^0 \alpha).$$

There are no other periodic points of S. Let $\theta : \{0, 1, 2, 3\} \to \{0, 1\}$ be defined by $\theta(0) = \theta(1) = 1$ and $\theta(2) = \theta(3) = 0$, then the restriction of the sequence $\hat{\theta}(\underline{f}_{(3,0)})$ on $\mathbb{N} = \mathbb{Z}_e$ is called the Baum–Sweet sequence. It is recursively defined by

$$\underline{bs}(x^0) = 1$$
$$\underline{bs}(x^{2j+1}) = \underline{bs}(x^j)$$
$$\underline{bs}(x^{4j}) = \underline{bs}(x^j)$$
$$\underline{bs}(x^{4j+2}) = 0.$$

By the recursive definition of $\underline{bs}$, it is easy to see that $\underline{bs}(x^j) = 1$ if and only if $j \geq 0$ and the binary expansion of j contains no substring of the form $10\ldots0*$[1] such that the number of zeros is odd.

3. Let $\overline{\mathcal{A}} = \{0, 1, 2, 3\}$, $\Gamma = \langle x \rangle$, $H(x^j) = x^{2j}$ and $V = \{x^0, x^1\}$ and define $s : V \times \mathcal{A} \to \mathcal{A}$ by

	x^0	x^1
0	0	1
1	2	1
2	0	3
3	2	3

The induced (V, H)-substitution S on $\Sigma(\mathbb{Z}, \mathcal{A})$ has six fixed points and no other periodic points. If θ is defined as in 2. above, then the sequence

$$\underline{pf} = \hat{\theta}(\lim_{n \to \infty} S^{\circ n}(x^0 0)),$$

regarded as a sequence in $\Sigma(\mathbb{N}, \{0, 1\})$, is called the paperfolding sequence. By construction, the paperfolding sequence is recursively defined by

$$\underline{pf}(x^{2j+1}) = \underline{pf}(x^j)$$
$$\underline{pf}(x^{4j}) = 1$$
$$\underline{pf}(x^{4j+2}) = 0,$$

for $j \geq 0$.

4. Let $H(x^j) = x^{3j}$ and let $V = \{x^{-1}, x^0, x^1\}$ and define $s : V \times \mathcal{A} \to \mathcal{A}$ by

	x^{-1}	x^0	x^1
0	2	0	1
1	0	1	2
2	2	2	2

---

[1] The symbol $*$ is either 1 or the end of the binary expansion

Since V is a complete digit set, the number of periodic points of the induced substitution S is equal to the number of periodic points of the map $s_{x^0} : \mathcal{A} \to \mathcal{A}$ which is 3. All periodic points of S are fixed points of S. We denote the fixed points by

$$\underline{f}_j = \lim_{n \to \infty} S^{\circ n}(x^0 \, j),$$

where $j \in \{0, 1\}$.

Let $\theta : \{0, 1, 2\} \to \{0, 1\}$ be defined by $\theta(0) = 1$ and $\theta(1) = \theta(2) = 0$. The restriction of the sequence $\hat{\theta}(\underline{f}_0)$ on $\mathbb{N}$ is abbreviated as $\underline{cs}$, and we have $cs(x^j) = 1$ if and only if the 3-adic expansion of j contains no 1, i.e., $j = \sum_{i=0}^{\infty} 3^i \, n_i$, where $n_i \in \{0, 2\}$ for all i.

5. Let $\Gamma = \langle x, y \rangle$ be the free Abelian group generated by two elements, i.e, $\Gamma = \mathbb{Z}^2$, and let $H(x^i y^j) = x^{3i} y^{3j}$. Then H is expanding (w.r.t. $E = \{x, y\}$). Let $\mathcal{A} = \{0, 1, 2, 3\}$ and $V = \{x^\alpha y^\beta \mid \alpha, \beta = 0, 1, 2\}$, then the sequence

$$\underline{b}(x^i y^j) = \begin{cases} \varnothing & \text{if } i < 0 \text{ or } j < 0 \\ \binom{i+j}{i} \bmod 3 & \text{otherwise} \end{cases}$$

is generated by a (V, H)-substitution. This sequence is the sequence of the binomial coefficients modulo three. The substitution is given by a map $s : V \times \mathcal{A} \to \mathcal{A}$, which is defined by

$$s(x^\alpha y^\beta, a) = \begin{cases} \varnothing & \text{if } a = \varnothing \\ a \binom{\alpha+\beta}{\alpha} \bmod 3 & \text{otherwise,} \end{cases}$$

where we consider $a \in \overline{\mathcal{A}}, a \neq \varnothing$ as an element of the natural numbers. Note that the induced substitution $S : \Sigma(\mathbb{Z}^2, \{0, 1, 2\}) \to \Sigma(\mathbb{Z}^2, \{0, 1, 2\})$ has 81 different fixed points and no periodic points.

Note that $\mathbb{N}^2 = \mathbb{Z}^2_{x^0 y^0}$ is (V, H)-substitution-invariant. If we restrict S to $\Sigma(\mathbb{N}^2, \{0, 1, 2\})$, then S has only three fixed points.

The introduction of the decimation maps which are special reductions (Definitions 1.1.1), will enable us to develop a necessary and sufficient condition for a sequence $\underline{f}$ to be generated by a substitution.

Definition 2.2.12. Let $H : (\Gamma, \| \; \|) \to (\Gamma, \| \; \|)$ be an expanding endomorphism and let $\tau \in \Gamma$, then the map

$$\partial_\tau^H : \Sigma(\Gamma, \mathcal{A}) \to \Sigma(\Gamma, \mathcal{A})$$
$$\underline{f} \mapsto (T_\tau \circ H)^*(\underline{f})$$

is called the τ-*decimation* w.r.t. H.

Remarks.

1. In the notion of formal series the τ-decimation is written as

$$\partial_\tau^H(\underline{f}) = \bigoplus \gamma f_{\tau H(\gamma)} = \bigoplus \gamma \, \underline{f}(vH(\gamma)).$$

In other words, a τ-decimation is a $(T_\tau \circ H)$-reduction.

2. We have the following identity

$$(T_\tau \circ H)_* \circ \partial_\tau^H(\underline{f})(\gamma) = \begin{cases} \underline{f}(\gamma) & \text{if } \gamma = \tau H(\gamma') \\ \varnothing & \text{otherwise.} \end{cases}$$

3. Let $a : \mathcal{A} \to \mathcal{A}$ be any map and let $\hat{a} : \Sigma(\Gamma, \mathcal{A}) \to \Sigma(\Gamma, \mathcal{A})$ be the induced map, then

$$\partial_\tau^H \circ \hat{a} = \hat{a} \circ \partial_\tau^H$$

holds for all $\tau \in \Gamma$.

4. $\Sigma(\Gamma, \mathcal{A})$ regarded as a subset of $\Sigma(\Gamma, \overline{\mathcal{A}})$ is invariant under τ-decimations.

Corollary 2.2.13. *Let $H : (\Gamma, \| \ \|) \to (\Gamma, \| \ \|)$ be an expanding endomorphism, and let $\tau, \rho \in \Gamma$, then we have*

1. $\partial_\tau^H \circ \partial_\rho^H = \partial_{\rho H(\tau)}^{H^{\circ 2}},$

2. $\partial_\tau^H \circ (T_\rho)^* = \partial_{\rho\tau}^H,$

3. $\partial_\tau^H \circ (T_\rho)_* = \partial_{\rho^{-1}\tau}^H.$

Proof. 1.

$$\begin{aligned}
\partial_\tau^H \circ \partial_\rho^H &= (T_\tau \circ H)^* \circ (T_\rho \circ H)^* \\
&= (T_\rho \circ H \circ T_\tau \circ H)^* \\
&= (T_\rho \circ T_{H(\tau)} \circ H^{\circ 2})^* \\
&= (T_{\rho H(\tau)} \circ H^{\circ 2})^* \\
&= \partial_{\rho H(\tau)}^{H^{\circ 2}}.
\end{aligned}$$

2. $\partial_\tau^H \circ (T_\rho)^* = (T_\rho \circ T_\tau \circ H)^* = \partial_{\rho\tau}^H.$

3. Since the map $T_\rho : \Gamma \to \Gamma$ is bijective we have $(T_\rho)_* = (T_{\rho^{-1}})^*$ and the assertion follows from 2. $\qquad\square$

Lemma 2.2.14. *Let $H : (\Gamma, \| \ \|) \to (\Gamma, \| \ \|)$ be an expanding endomorphism such that $H(\Gamma)$ is a subgroup of finite index. If V is a residue set of H, then any τ-decimation ∂_τ^H can be written as*

$$\partial_\tau^H = (T_{\kappa(\tau)})^* \circ \partial_{\zeta(\tau)}^H,$$

where κ and ζ are the associated image-part- and remainder-map, respectively.

Proof. Any $\tau \in \Gamma$ has a unique representation as $\tau = \zeta(\tau) H(\kappa(\tau)) = vH(\tau')$. We therefore obtain that

$$\begin{aligned}
\partial_\tau^H = \partial_{vH(\tau')}^H &= (T_{vH(\tau')} \circ H)^* \\
&= (T_v \circ H \circ T_{\tau'})^* \\
&= (T_{\tau'})^* \circ \partial_v^H,
\end{aligned}$$

which finishes the proof. □

A combination of Lemma 2.2.14 and 2. of Corollary 2.2.13 yields

Lemma 2.2.15. *Let H be an expanding endomorphism of Γ, let V be a residue system of H, and let ζ and κ be the associated remainder-map and image-part-map, respectively. Then*

$$\partial_v^H \circ (T_\rho)^* = (T_{\kappa(\rho v)})^* \circ \partial_{\zeta(\rho v)}^H$$

holds for all $\rho \in \Gamma$ and $v \in V$.

The next lemma provides a kind of summation formula for elements of $\Sigma(\Gamma, \overline{\mathcal{A}})$.

Lemma 2.2.16. *Let the assumptions of Lemma 2.2.15 be satisfied. Then*

$$\underline{f} = \bigoplus_{v \in V} (T_v \circ H)_* \circ \partial_v^H(\underline{f})$$

holds for all $\underline{f} \in \Sigma(\Gamma, \overline{\mathcal{A}})$.

The proof is a direct consequence of Remark 2, p. 39.

The following lemma relates v-decimations with $v \in V$, V a residue set, to (V, H)-substitutions.

Lemma 2.2.17. *Let the assumptions of Lemma 2.2.15 be satisfied, and let $S : \Sigma(\Gamma, \mathcal{A}) \to \Sigma(\Gamma, \mathcal{A})$ be a (V, H)-substitution induced by a map $s : V \times \mathcal{A} \to \mathcal{A}$. Then*

$$\partial_v^H \circ S(\underline{f}) = \widehat{s_v}(\underline{f})$$

holds for all $v \in V$ and $\underline{f} \in \Sigma(\Gamma, \mathcal{A})$.

Proof. By Definition 2.2.8, we can write $S(\underline{f})$ as

$$S(\underline{f}) = \bigoplus_{w \in V} (T_w \circ H)_* \circ \widehat{s_w}(\underline{f}),$$

by Corollary 1.1.2 and by Lemma 2.2.16, we obtain $\partial_v^H(S(\underline{f})) = \widehat{s_v}(\underline{f})$. $\qquad\square$

Remark. If Γ' is a (V, H)-substitution-invariant subset of Γ, then the v-decimations, $v \in V$ are well-defined maps from $\Sigma(\Gamma', \mathcal{A}) \to \Sigma(\Gamma', \overline{\mathcal{A}})$.

The most important object which allows us to describe sequences generated by a substitution, is the kernel of a sequence.

Definition 2.2.18. Let $H : \Gamma \to \Gamma$ be an expanding endomorphism and let V denote a residue set of H. For any $\underline{f} \in \Sigma(\Gamma, \mathcal{A})$ the set

$$\{\underline{f}\} \cup \{\partial_{v_1}^H \circ \cdots \circ \partial_{v_n}^H(\underline{f}) \mid v_i \in V, n \in \mathbb{N}\}$$

is called the (V, H)-*kernel* of $\underline{f}$ and is denoted by $\ker_{V,H}(\underline{f})$.

Remark. If Γ' is a (V, H)-substitution-invariant set, then the (V, H)-kernel of the sequence $\underline{f} \in \Sigma(\Gamma', \mathcal{A})$ is well defined.

Examples.
1. The (V, H)-kernel of the Thue–Morse sequence, viewed as an element of $\Sigma(\Gamma_e, \{0, 1\})$, consists of the elements $\underline{t} = (t_n)$ and $\underline{u} = (\overline{t}_n)$, where $\overline{0} = 1$ and $\overline{1} = 1$.

2. The (V, H)-kernel of the Baum–Sweet sequence $\underline{bs}$, cf., Example 2, p. 36, contains three elements which are $\underline{bs}$, $\oplus x^j \underline{bf}(x^{2j})$ and $\oplus_{j \geq 0} x^j \, 0$. This follows almost immediately from the recursive description of $\underline{bf}$.

3. The (V, H)-kernel of the paperfolding sequence contains $\underline{pf}$, the sequence $\oplus x^j \underline{pf}(x^{2j})$, and the sequences $\oplus_{j \geq 0} x^j 1, \oplus_{j \geq 0} x^j 0$.

4. The (V, H)-kernel of the binomials modulo three contains the sequences $\underline{b}$, $\oplus_{i,j \geq 0} x^i y^j 2\underline{b}(x^i y^j)$ mod 3 and $\oplus_{i,j \geq 0} x^i y^j 0$.

The next theorem states a necessary and sufficient condition for a sequence to be generated by a (V, H)-substitution.

Theorem 2.2.19. *Let $\underline{f} \in \Sigma(\Gamma, \mathcal{A})$. Then $\underline{f}$ is generated by a (V, H)-substitution if and only if the (V, H)-kernel of $\underline{f}$ is a finite set.*

Proof. Let us assume that f is generated by a (V, H)-substitution, i.e., $f = \hat{\theta}(\underline{F})$, where $\underline{F}$ is a fixed point of a (V, H)-substitution $S : \Sigma(\Gamma, \mathcal{B}) \to \Sigma(\Gamma, \mathcal{B})$. We have $\partial_v^H \circ \hat{\theta} = \hat{\theta} \circ \partial_v^H$, therefore it suffices to show that $\ker_{V,H}(\underline{F})$ is finite.

Since $S(\underline{F}) = \underline{F}$ and by Lemma 2.2.17, we obtain

$$\partial_v^H(\underline{F}) = \partial_v^H(S(\underline{F})) = \widehat{s_v}(\underline{F}).$$

This yields $\partial_w^H \circ \partial_v^H(S(\underline{F})) = \partial_w^H(\widehat{s_v}(\underline{F})) = \widehat{s_v} \circ \partial_w^H(\underline{F}) = \widehat{s_v} \circ \widehat{s_w}(\underline{F})$. Thus, we obtain

$$\ker_{V,H}(\underline{F}) = \{\underline{F}\} \cup \{\widehat{s_{v_1}} \circ \cdots \circ \widehat{s_{v_n}}(\underline{F}) \mid v_j \in V; j = 1, \ldots, n; n \in \mathbb{N}\}.$$

We conclude that the (V, H)-kernel of $\underline{F}$ is a subset of the set

$$\overline{K} = \{\hat{a}(\underline{F}) \mid \text{where } a \text{ is any map } a : \mathcal{B} \to \mathcal{B}\}.$$

The above set is finite since $\mathcal{B}$ is finite, therefore the (V, H)-kernel of $\underline{F}$ is finite.

Now, let us assume that the (V, H)-kernel of f is finite, i.e.,

$$\ker_{V,H}(\underline{f}) = \{\underline{f}_1, \ldots, \underline{f}_N\},$$

where $N \in \mathbb{N}$ and $\underline{f}_1 = \underline{f}$. As a consequence of the finiteness of the (V, H)-kernel of $\underline{f}$ there exists a map $v : V \times \{1, 2, \ldots, N\} \to \{1, \ldots, N\}$ such that $\partial_v^H(\underline{f}_j) = \underline{f}_{v(v,j)}$ for all $v \in V$ and $j \in \{1, \ldots, N\}$. Due to Lemma 2.2.16 we obtain

$$\underline{f}_j = \bigoplus_{v \in V}(T_v \circ H)_* \circ \partial_v^H(\underline{f}_j) = \bigoplus_{v \in V}(T_v \circ H)_*(\underline{f}_{v(v,j)}) \qquad (2.2)$$

for all $j \in \{1, \ldots, N\}$. Equation (2.2) can be interpreted as a (V, H)-substitution in the following way. Define $\mathcal{B} = \mathcal{A}^N$. Then the map

$$s : V \times \mathcal{B} \to \mathcal{B}, \quad s(v, (a_1, \ldots, a_N)) = (a_{v(v,1)}, \ldots, a_{v(v,N)})$$

induces a substitution $S : \Sigma(\Gamma, \mathcal{B}) \to \Sigma(\Gamma, \mathcal{B})$. Let $\underline{F} \in \Sigma(\Gamma, \mathcal{B})$ be defined by

$$\underline{F}(\gamma) = (\underline{f}_1(\gamma), \ldots, \underline{f}_N(\gamma)).$$

Due to Equation (2.2), the above defined $\underline{F}$ is a fixed point of the substitution S. If we define $\theta : \mathcal{B} \to \mathcal{A}$ as $\theta((a_1, \ldots, a_N)) = a_1$, we obtain

$$\hat{\theta}(\underline{F}) = \underline{f}_1 = \underline{f}.$$

This finishes the proof. □

The first part of the proof of Theorem 2.2.19 is a fundamental tool to establish the finiteness of the kernel of a given sequence $\underline{f}$. The procedure is as follows: Find a finite set $\overline{K}$ which contains $\underline{f}$ and is invariant under all v-decimations, then $\ker_{V,H}(\underline{f})$ being a subset of $\overline{K}$ is finite. We shall encounter this method of proof several times in the following chapters.

As the proof of Theorem 2.2.19 shows, we can compute the kernel of a sequence which is a fixed point of a substitution.

Example. Let $\underline{t}$ be the Thue–Morse sequence. Then we have

$$\partial_{x^0}^H(\underline{t}) = \hat{s}_{x_0}(\underline{t}) = \underline{t}$$

and

$$\partial_{x^1}^H(\underline{t}) = \hat{s}_{x_1}(\underline{t}) = \underline{u}.$$

Thus we get

$$\partial_{x^0}(\partial_{x^0}^H(\underline{t})) = \hat{s}_{x_0}(\hat{s}_{x_1}(\underline{t})) = \underline{u},$$
$$\partial_{x^1}(\partial_{x^0}^H(\underline{t})) = \hat{s}_{x_1}(\hat{s}_{x_1}(\underline{t})) = \underline{u}.$$

Therefore the (V, H)-kernel of the Thue–Morse sequence is the set $\{\underline{t}, \underline{u}\}$.

In order to compute the (V, H)-kernel of a fixed point $\underline{F}$ of a (V, H)-substitution we introduce the graph of a substitution. Let S be a (V, H)-substitution and $s_v, v \in V$ be its defining maps. The maps s_v and $\mathrm{id}_{\mathcal{A}}$ generate via composition a finite semigroup, $G = G(S)$ which is a subset of all maps from $\mathcal{A}$ to $\mathcal{A}$.

A finite directed labeled graph with base point is a quadruple (E, b, L, K). E denotes the set of vertices, $b \in E$ is the base point, L denotes the set of labels and K denotes the set of directed edges. An edge is denoted by $g \xrightarrow{l} h$, where $g, h \in E$ and $l \in L$. Moreover, all sets are finite.

Definition 2.2.20. Let $s : V \times \mathcal{A} \to \mathcal{A}$ be a (V, H)-substitution. The (V, H)-*substitution-graph* is the directed, labeled graph with base point defined by $(G(S), \mathrm{id}_{\mathcal{A}}, \{s_v \mid v \in V\}, K)$, i.e., the set of vertices is the semigroup $G(S)$ of the substitution, the basepoint is $\mathrm{id}_{\mathcal{A}}$, the set of labels is $\{s_v \mid v \in V\}$, and the set of labeled edges K is defined as follows: If $g_1, g_2 \in G(S)$ and $v \in V$, then $(g_1 \xrightarrow{s_v} g_2) \in K$ if and only if $s_v \circ g_1 = g_2$.

Remarks.

1. The (V, H)-substitution-graph can be considered as a Cayley-graph of the semigroup $G(S) = \langle \{s_v \mid v \in V\} \rangle$. To be precise, it should be called left Cayley graph, since an edge from g_1 to g_2 with label s_v exists if and only if $s_v \circ g_1 = g_2$. Later we shall see that the right Cayley graph of $G(S)$ is also meaningful for sequences generated by a substitution.

2. Fix $b = a_1 a_2 \ldots a_N \in \mathcal{A}^N$, $N = |\mathcal{A}|$, such that $a_i \neq a_j$ whenever $i \neq j$ and let $g \in G(S)$, then we can encode the elements of $G(S)$ by its action on b. For a given $g \in G(S)$, we write $g(b) = g(a_1)g(a_2)\ldots g(a_N)$. Thus we can describe g as $g(a_1)\ldots g(a_N) \in \mathcal{A}^N$. In other words the vertices of the (V, H)-substitution-graph are given by $g(a_1) \quad g(a_N) \in \mathcal{A}^N$ and b denotes the basepoint of the (V, H)-substitution-graph.

3. Let f be a fixed point of a (V, H)-substitution S and $G(S)$ the associated semigroup, then $\ker_{V,H}(\underline{f}) \subset \{\hat{g}(\underline{f}) \mid g \in G(S)\}$.

4. If $\underline{F}$ is a fixed point of a (V, H)-substitution S and $\theta : \mathcal{B} \rightarrow \mathcal{A}$ a map, then the (V, H)-kernel of $\hat{\theta}(\underline{F})$ is contained in the set $\{\hat{\theta}(\hat{g}(\underline{f})) \mid g \in G(S)\}$. Moreover, if $\theta \circ g = \theta \circ h$ for $g, h \in G(S)$, then $\hat{\theta}(\hat{g}(\underline{F})) = \hat{\theta}(\hat{h}(\underline{F}))$. Thus, the cardinality of the set $\{\theta \circ g \mid g \in G(S)\}$ is an upper bound for the cardinality of $\ker_{V,H}(\hat{\theta}(\underline{F}))$.

Examples.

1. Let $\underline{t}$ be the Thue–Morse sequence, then the (V, H)-substitution-graph is given in Figure 2.1. The double circle represents the base point. This will be the case in all figures!

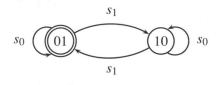

Figure 2.1. The substitution-graph of the Thue–Morse substitution.

Each vertex of the graph represents an element of the semigroup of the substitution. Therefore the kernel of $\underline{t}$ contains two elements, as we have seen earlier.

2. Let $\underline{bs} = \hat{\theta}(\underline{f}_{(3,0)})$ be the Baum–Sweet sequence, see Example 2, p. 36. The graph of the associated substitution is given in Figure 2.2.

Therefore the (V, H)-kernel of $\underline{f}_{(3,0)}$ contains at most 7 elements. By inspection, one can see that the kernel of $\underline{f}_{(3,0)}$ contains indeed seven different elements.

Since $\theta \circ (0123) = \theta \circ (1133)$, $\theta \circ (0213) = \theta \circ (1213)$, and $\theta \circ (2233) = \theta \circ (2323) = \theta \circ (3333)$, the kernel of the Baum–Sweet sequence contains at most three elements. In fact, as one can easily see the kernel of the Baum–Sweet sequence contains three elements.

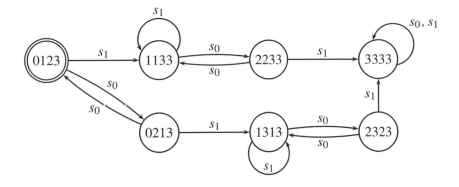

Figure 2.2. The substitution-graph of the Baum–Sweet substitution.

3. Let $\mathcal{A} = \{0, 1, 2, 3\}$, $\Gamma = \langle x \rangle = \mathbb{Z}$, $H(x^j) = x^{2j}$ and $V = \{x^0, x^1\}$ then $s : V \times \mathcal{A} \to \mathcal{A}$ defined by

	x^0	x^1
0	0	1
1	1	2
2	2	0
3	2	3

induces a (V, H)-substitution S on $\Sigma(\mathbb{Z}, \mathcal{A})$. Let

$$\underline{F}_{(3,0)} = \lim_{n \to \infty} S^{\circ n}(X^{-1} 3 \oplus x^0 0)$$

be one of the fixed points of the substitution S. The (V, H)-substitution-graph is given in Figure 2.3. The number of kernel elements is therefore limited by 13. The actual value of the number of elements is three.

As the above example shows, the number of elements of $G(S)$ gives only an upper bound for the cardinality of the kernel. We next show how to determine the cardinality of $\ker_{V,H}(\underline{f})$, where $\underline{f}$ is a fixed point of a substitution.

Definition 2.2.21. Let $\underline{f} \in \Sigma(\Gamma, \mathcal{A})$. The set

$$\mathcal{R}(\underline{f}) = \{a \mid \text{there exists } \gamma \in \Gamma \text{ such that } \underline{f}(\gamma) = a\}$$

is called the *range* of $\underline{f}$.

Remarks.

1. If $\underline{f}$ is a fixed point of a substitution S, then $\mathcal{R}(\underline{f})$ is $G(S)$-invariant, i.e., $g(\overline{\mathcal{R}(\underline{f})}) \subset \mathcal{R}(\underline{f})$ holds for all $g \in G(S)$.

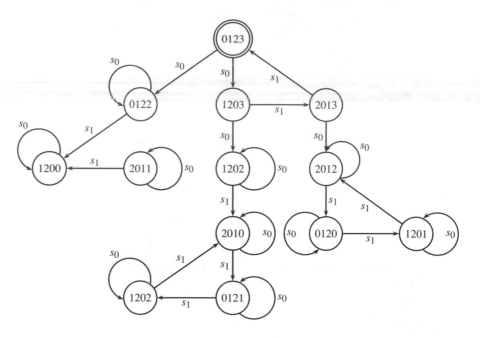

Figure 2.3. The substitution-graph of the substitution of Example 3, p. 45.

2. The range $\mathcal{R}(f)$ of a fixed point f is the smallest $G(S)$-invariant subset of $\overline{\mathcal{A}}$ that contains the set $\{\underline{f}(\gamma) \mid \gamma \in \overline{\mathrm{Per}}\, \kappa_{V,H}\}$.

3. Let $g, h \in G(S)$ be elements of the semigroup associated with the (V, H)-substitution S and let f be a fixed point of S. We say that g is $\mathcal{R}(f)$-equivalent to h if $g_{|\mathcal{R}(f)} = h_{|\mathcal{R}(f)}$. The quotient of $G(S)$ w.r.t. $\mathcal{R}(f)$-equivalence is a semigroup. This follows from the fact that $g(\mathcal{R}(f)) \subset \mathcal{R}(f)$. We denote the quotient by $G_{\mathcal{R}(f)}(S)$.

Examples.

1. The range of the Thue–Morse sequence is $\mathcal{A} = \{0, 1\}$.

2. The range of the sequence $\underline{f}_{(3,0)}$ which generates the Baum–Sweet sequence, is $\mathcal{A} = \{0, 1, 2, 3\}$.

3. The range of the sequence $\underline{F}_0$, Example 3. above, is $\{0, 1, 2\}$. The quotient of $G(S)$ modulo $\mathcal{R}(\underline{F}_0)$-equivalence is the semigroup

$$G_{\mathcal{R}(\underline{F}_{(3,0)})}(S) = \{(012), (120), (021)\},$$

where we use the notation for maps on the finite set $\mathcal{R}(\underline{F}_{(3,0)})$.

Lemma 2.2.22. *Let $\underline{F}$ be a fixed point of the (V, H)-substitution S, then*

$$\left| \ker_{V,H}(\underline{F}) \right| = \left| G_{\mathcal{R}(\underline{F})}(S) \right|.$$

Proof. The map $g \mapsto \hat{g}(\underline{F})$ defines a surjection from $G_{\mathcal{R}(\underline{F})}(S)$ to $\ker_{V,H}(\underline{F})$. Therefore $|G_{\mathcal{R}(\underline{F})}(S)| \geq |\ker_{V,H}(\underline{F})|$.

Now suppose that there are $g, h \in G_{\mathcal{R}(\underline{F})}(S)$ such that $g \neq h$ and $\hat{g}(\underline{F}) = \hat{h}(\underline{F})$. Since $g \neq h$ there exists an $a \in \mathcal{R}(\underline{F})$ such that $g(a) \neq h(a)$; but this contradicts $\hat{g}(\underline{F}) = \hat{h}(\underline{F})$. □

Corollary 2.2.23. *Let $\underline{F} \in \Sigma(\Gamma, \mathcal{B})$ be a fixed point of the (V, H)-substitution S and let $\theta : \mathcal{B} \to \mathcal{A}$ be a map. Then $\hat{\theta}(\underline{F})$ is generated by a substitution and we have*

$$\left| \ker_{V,H}(\hat{\theta}(\underline{F})) \right| = \left| \{\theta \circ g \mid g \in G_{\mathcal{R}(\underline{F})}(S)\} \right|.$$

The proof follows the same lines as the proof of the above lemma.

Definition 2.2.24. Let $\underline{F}$ be a fixed point of a (V, H)-substitution S. The $\underline{F}$-*graph* of the (V, H)-substitution is a directed, labelled graph defined as follows: The set of vertices is the semigroup $G_{\mathcal{R}(\underline{F})}(S)$, and the set of labels $\{s_v \mid v \in V\}$. There is a directed edge with label s_v from g_1 to g_2 if $s_v \circ g_1 = g_2$.

The vertex $\mathrm{id}_{\mathcal{R}(\underline{F})}$ is called the *basepoint* of the $\underline{F}$-graph.

Remark. The $\underline{F}$-graph is the left Cauchy graph of the semigroup $G_{\mathcal{R}(\underline{F})}(S)$.

Examples.
1. Consider Example 3 from above. The semigroup $G_{\mathcal{R}(\underline{F}_{(3,0)})}(S)$ contains three elements which are $(012), (120), (021)$. The $\underline{F}_{(3,0)}$-graph is shown in Figure 2.4.

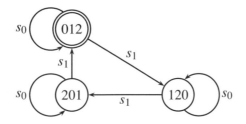

Figure 2.4. The $\underline{F}_{(3,0)}$-graph of the substitution of Example 3.

2. Let pf be the paper folding sequence, considered as an element of $\Sigma(\Gamma_e, \{0, 1\})$, see Example 3, p. 37, and let $\underline{f}_0 = \lim_{n \to \infty} S^{\circ n}(x^0\, 0)$ be considered as an element of $\Sigma(\Gamma_e, \{0, 1, 2, 3\})$. Then we have $\hat{\theta}(\underline{f}_0) = \underline{pf}$. The semigroup $G(S)$

of the substitution S contains 7 elements. Since the range of $\underline{f}_0$ is $\{0, 1, 2, 3, \varnothing\}$, the cardinality of the kernel of $\underline{f}_0$ is equal to 7, see Lemma 2.2.22. By Corollary 2.2.23, the cardinality of the kernel of the paper folding sequence is equal to $|\{\theta \circ g \mid g \in G_{\mathscr{R}(f)}(S)\}|$, which is 4.

In the preceding we have seen that the left Cauchy graph of the semigroup resembles certain features of the substitution. Moreover, by factoring out certain elements of $G(S)$ we constructed the semigroup $G_{\mathscr{R}(F)}(S)$ which provides us with information about the fixed point $\underline{F}$ of the substitution. Even if we consider a sequence of the form $\hat{\theta}(\underline{f})$ the above concepts are useful, see Corollary 2.2.23.

Definition 2.2.25. Let $\underline{f}$ be generated by a (V, H)-substitution. Then the (V, H)-*kernel graph* of $\underline{f}$ is a directed, labeled graph defined as follows: The set of vertices is the (V, H)-kernel of $\underline{f}$, and the set of labels is the set V. There is a directed edge with label v from $\underline{f}_1$ to $\underline{f}_2$ if and only if $\partial_v^H(\underline{f}_1) = \underline{f_2}$.
The vertex $\underline{f}$ is called the *basepoint* of the (V, H)-kernel graph.

Remark. If $M = \{\underline{f}_1, \ldots, \underline{f}_N\}$ is a finite set such that $\partial_v^H(M) \subset M$ for all $v \in V$, then the (V, H)-kernel-graph of the set M is a directed graph with set of vertices equal to M, label set equal to V, and the set of edges is given by $\{(\underline{f} \xrightarrow{v} \underline{h}) \mid \underline{f}, \underline{h} \in M, \ \partial_v^H(\underline{f}) = \underline{h}\}$. If we choose a basepoint $\underline{f}_j \in M$ the (V, H)-kernel-graph of M with basepoint $\underline{f}_j$ is called extended (V, H)-kernel-graph of $\underline{f}_j$.

Examples.

1. Let $\underline{t}$ be the Thue–Morse sequence, its kernel graph is shown in Figure 2.5, where 0 and 1 are abbreviations for x^0, x^1, respectively.

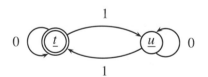

Figure 2.5. The kernel-graph of the Thue–Morse sequence.

2. In Figure 2.6 we see the kernel graph of the Baum–Sweet sequence. As for the Thue–Morse sequence, 0 and 1 stand for x^0 and x^1. Furthermore, $\underline{f} = \oplus_{j \geq 0} x^j \, \underline{bs}(x^{2j})$ and $\underline{h} = \oplus_{j \geq 0} x^j \, 0$.

The next theorem shows how to compute the kernel graph of a sequence that is generated by a substitution.

To this end we introduce the right Cayley graph of the semigroup $G(S)$ of a substitution S. The right Cayley graph is a directed, labeled graph. Its vertices are the

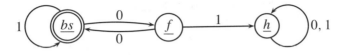

Figure 2.6. The kernel-graph of the Baum–Sweet sequence.

elements of $G(S)$, the labels are the maps s_v and for $g_1, g_2 \in G(S)$ and $v \in V$ we have that $g_1 \xrightarrow{s_v} g_2$ is a labeled edge of the right Cayley graph if and only if $g_1 \circ s_v = g_2$.

Moreover, we need the concept of isomorphic directed, labeled graphs with a base point. Let (E, b, L, K) be any finite, directed, labeled graph, where E denotes the set of vertices, b the basepoint, L the labels and K the arrows, denoted by $(a_1 \xrightarrow{l} a_2)$. Let $\vartheta : E \to E'$ be a surjective map from E to a finite set E'. The ϑ-projection of the graph (E, b, L, K) is the directed graph $(E', \vartheta(b), L, K')$, where

$$K' = \{(a_1' \xrightarrow{l} a_2') \mid \text{there exist } (a_1 \xrightarrow{l} a_2) \in K \text{ such that } \vartheta(a_1) = a_1', \ \vartheta(a_2) = a_2'\}$$

is the set of edges, and $\vartheta(b)$ is the basepoint.

Two directed, labeled graphs $(E, b, L, K), (E', b', L', K')$ are isomorphic if there exist bijective maps $\vartheta : E \to E'$, $\lambda : L \to L'$ such that $\vartheta(b) = b'$ and the map $(\vartheta, \lambda) : K \to K'$ given by

$$(\vartheta, \lambda)((a_1 \xrightarrow{l} a_2)) = (\vartheta(a_1) \xrightarrow{\lambda(l)} \vartheta(a_2))$$

is well defined and bijective.

Theorem 2.2.26. *Let $\underline{F}$ be a fixed point of the (V, H)-substitution S. Then the (V, H)-kernel-graph of $\underline{F}$ is isomorphic to the right Cayley graph of $G_{\mathcal{R}(\underline{F})}(S)$.*

Proof. By Lemma 2.2.22, we have

$$\left| \ker_{V,H}(\underline{F}) \right| = \left| G_{\mathcal{R}(\underline{F})}(S) \right|$$

and the map $\Delta : G_{\mathcal{R}(\underline{F})}(S) \to \ker_{V,H}(\underline{F})$ defined by $\Delta(g) = \hat{g}(\underline{F})$ is a bijection. Define the map $\lambda : \{s_v \mid v \in V\} \to V$ by $\lambda(s_v) = v$. Certainly, λ is a bijection. Then the pair (Δ, λ) is a bijection between the right Cayley graph of $G_{\mathcal{R}(\underline{F})}(S)$ and the (V, H)-kernel-graph of $\underline{F}$.

It remains to show that (Δ, λ) is a bijection of the edges. To this end it suffices to prove that (Δ, λ) is well defined.

Let $(g_1 \xrightarrow{s_v} g_2)$ be a directed edge of the Cayley graph of $G_{\mathcal{R}(\underline{F})}(S)$, then we have

$$(\Delta, \lambda)((g_1 \xrightarrow{s_v} g_2)) = (\hat{g}_1(\underline{F}) \xrightarrow{v} \hat{g}_2(\underline{F})).$$

Therefore we must show that $(\hat{g}_1(\underline{F}) \xrightarrow{v} \hat{g}_2(\underline{F}))$ is an edge of the kernel graph of f, i.e., we have to prove that $\partial_v^H(\hat{g}_1(\underline{F})) = \hat{g}_2(\underline{F})$.

By Remark 3, p. 39, we have

$$\partial_v^H(\hat{g}_1(\underline{F})) - \hat{g}_1(\partial_v^H(\underline{F})).$$

By Lemma 2.2.17, we obtain

$$\hat{g}_1(\partial_v^H(\underline{F})) = \hat{g}_1(s_v(\underline{F})) = \hat{g}_2(\underline{F}).$$

Thus, we have proved that (Δ, λ) maps arrows of the Cayley graph on arrows of the kernel graph of f.

The bijectivity property of the map (Δ, λ) on the edges is then an immediate consequence of the bijectivity property of Δ and λ.   □

Corollary 2.2.27. *Let $\underline{F} \in \Sigma(\Gamma, \mathcal{B})$ be a fixed point of the (V, H)-substitution S and let $\theta : \mathcal{B} \to \mathcal{A}$ be a map. If $\underline{f} = \hat{\theta}(\underline{F})$ and if $\tilde{\theta} : G_{\mathcal{R}(\underline{F})}(S) \to \ker_{V,H}(\underline{f})$ is defined by $\tilde{\theta}(g) = \hat{\theta}(\hat{g}(\underline{F}))$, then the (V, H)-kernel graph of $\underline{f}$ is isomorphic to the $\tilde{\theta}$-projection of the right Cayley graph of $G_{\mathcal{R}(\underline{F})}(S)$.*

Proof. By Corollary 2.2.23, the map $\hat{\theta} : G_{\mathcal{R}(\underline{F})}(S) \to \ker_{V,H}(\underline{f})$ is bijective. Let $\lambda : \{s_v \mid v \in V\} \to V$ be defined as in the proof of Theorem 2.2.26. If $g_1 \xrightarrow{s_v} g_2$ is an edge in the right Cayley graph of $G_{\mathcal{R}(\underline{F})}(S)$, then it remains to show that $\partial_v^H(\tilde{\theta}(g_1)) = \tilde{\theta}(g_2)$. We have, cf. Remark 3, p. 39,

$$\partial_v^{(}\tilde{\theta}(g_1)) = \partial_v^H(\hat{\theta}(\hat{g}_1(\underline{F})))) = \hat{\theta}(\partial_v^H(\hat{g}_1(\underline{F}))).$$

By Lemma 2.2.17 this yields

$$\hat{\theta}(\partial_v^H(\hat{g}_1(\underline{F}))) = \hat{\theta}(\hat{g}_1(\hat{s}_v(\underline{F}))), = \tilde{\theta}(g_2)$$

which was our claim.   □

Example. We consider the paperfolding sequence $\underline{pf} = \theta(\lim_{n\to\infty} S^{\circ n}(x^0\ 0))$, see Example 2, p. 36. The range of $\underline{F} = \lim_{n\to\infty} S^{\circ n}(x^0\ 0)$ is $\{0, 1, 2, 3, \varnothing\}$. By Theorem 2.2.26, the kernel graph of $\underline{F}$ is isomorphic to the right Cayley graph of $G(S)$. The right Cayley graph is shown in Figure 2.7, where the labels $\{s_{x^0}, s_{x^1}\}$ are replaced by 0 and 1, respectively. Note that $G(S) = G_{\mathcal{R}(\underline{F})}(S)$ which implies that the $\tilde{\theta}$-projection of the left Cayley graph of $G(S)$ gives the kernel graph of $\underline{pf}$. In Figure 2.8 the θ-projection of the right Cayley graph is given. Due to Corollary 2.2.27, this is the kernel graph of $\underline{pf}$.

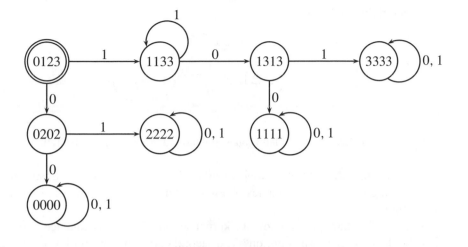

Figure 2.7. The right Cayley graph of the paperfolding substitution.

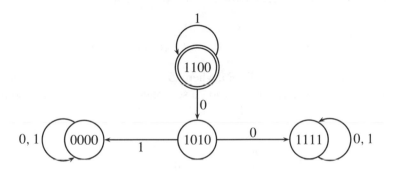

Figure 2.8. The kernel-graph of the paperfolding sequence.

Remarks.

1. The semigroup $G(S)$ of the paperfolding substitution is an example of the fact that the right Cayley graph and the left Cayley graph need not to be isomorphic as graphs. One easily computes that the left Cayley graph contains a cycle depicted in Figure 2.9, whereas the right Cayley graph has only trivial cycles, cf., Figure 2.7.

2. The right Cayley graph of the Thue–Morse substitution is isomorphic to the left Cayley graph. This is no surprise, since the semigroup $G(S)$ is in fact the commutative group $\mathbb{Z}/(2\mathbb{Z})$.

3. The Baum–Sweet substitution provides another example of non-isomorphic left

and right Cayley graphs.

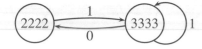

Figure 2.9. A cycle in left Cayley graph of the paperfolding substitution.

Let us summarize the preceding results. A knowledge of the substitution that generates a given sequence enables one to compute the kernel and the kernel graph quite efficiently.

On the other hand, if we know the kernel-graph of a given sequence what can be said about a generating substitution? The second part of the proof of Theorem 2.2.19 provides a method to compute a substitution of a sequence if its kernel graph is given.

Let $\underline{f} \in \Sigma(\Gamma, \mathcal{A})$ be such that $\ker_{V,H}(\underline{f}) = \{\underline{f}_j \mid j = 1, \ldots, N\}$ with $\underline{f}_1 = \underline{f}$. As the proof of Theorem 2.2.19 shows, there exists a map

$$s : V \times \mathcal{B} \to \mathcal{B}$$
$$s(v, (a_1, \ldots, a_N)) = (a_{\nu(v,1)}, \ldots, a_{\nu(v,N)})$$

that induces a substitution on $\Sigma(\Gamma, \mathcal{B})$, where $\mathcal{B} = \mathcal{A}^N$. This map s can be understood as a sum of certain matrices. In order to do so, we consider the field $\mathbb{F}_2$ and the finite set $\overline{\mathcal{A}} = \mathcal{A} \cup \{\varnothing\}$. For $b \in \mathbb{F}_2$ and $a \in \overline{\mathcal{A}}$ we define the product $a \cdot b$ by

$$b \cdot a = \begin{cases} a & \text{if } b = 1 \\ \varnothing & \text{if } b = 0. \end{cases}$$

For each $v \in V$ we define the matrix $A_v = (a_{ij}^v)_{i,j=1,\ldots,N} \in \mathbb{F}_2^{N \times N}$ by

$$a_{ij}^v = \begin{cases} 1 & \text{if } \nu(v, j) = i \\ 0 & \text{otherwise.} \end{cases}$$

In terms of the kernel graph of $\underline{f}$ we have $a_{ij}^v = 1$ if and only if $\partial_v^H(\underline{f}_i) = \underline{f}_j$. In other words, A_v is the adjacency matrix of the graph obtained from the kernel graph after removing all edges not labeled with v.

The maps $s_v : \mathcal{A}^N \to \mathcal{A}^N$ defined by the above substitution can then be written as a matrix product, i.e.,

$$s_v(a_1, \ldots, a_N) = A_v \begin{pmatrix} a_1 \\ \vdots \\ a_N \end{pmatrix},$$

where the sum is $\oplus$ on $\overline{\mathcal{A}}$, i.e, $a \oplus \varnothing = \varnothing \oplus a = a$ for all $a \in \overline{\mathcal{A}}$, and the products are as above. Since each row of A_v contains only one entry equal to 1, the above product is indeed well defined.

The substitution $S : \Sigma(\Gamma, \mathcal{A}^N) \to (\Gamma, \mathcal{A}^N)$ of the proof of Theorem 2.2.19 can now be written as

$$S(\underline{G}) = \bigoplus vH(\gamma)A_v(\underline{G}(\gamma)), \qquad (2.3)$$

where the summation is over all $v \in V$ and $\gamma \in \Gamma$.

On the other hand, if $A_v \in \mathbb{F}_2^{N \times N}$, $v \in V$, is a collection of matrices such that every row of every A_v contains precisely one 1, then Equation (2.3) defines a substitution.

Definition 2.2.28. Let $A_v \in \mathbb{F}_2^{N \times N}$, $v \in V$, such that every row of every matrix A_v contains one 1. The polynomial

$$\bigoplus_{v \in V} vA_v$$

is called a *substitution polynomial*.

If the matrices A_v are induced by a kernel graph of a sequence $\underline{f} \in \Sigma(\Gamma, \mathcal{A})$, then the substitution polynomial P is called the substitution polynomial of $\underline{f}$.

Examples.
1. Let $\underline{bs}$ be the Baum–Sweet sequence viewed as an element of $\Sigma(\mathbb{N}, \{0, 1\})$, see Example 2., p. 36. Its (V, H)-kernel-graph, where $V = \{x^0, x^1\}$, is shown in Figure 2.6. If we set $\underline{f}_1 = \underline{bf}$, $\underline{f}_2 = \underline{f}$, and $\underline{f}_3 = \underline{h}$, then the substitution polynomial is given by

$$x^0 \begin{pmatrix} 0 & 1 & 0 \\ 1 & 0 & 0 \\ 0 & 0 & 1 \end{pmatrix} \oplus x^1 \begin{pmatrix} 1 & 0 & 0 \\ 0 & 0 & 1 \\ 0 & 0 & 1 \end{pmatrix}.$$

2. The Thue–Morse sequence as a sequence in $\Sigma(\mathbb{N}, \{0, 1\})$ has the substitution polynomial for $V = \{x^0, x^1\}$.

$$x^0 \begin{pmatrix} 1 & 0 \\ 0 & 1 \end{pmatrix} \oplus x^1 \begin{pmatrix} 0 & 1 \\ 1 & 0 \end{pmatrix}.$$

3. The paper folding sequence as an element of $\Sigma(\Gamma, \{0, 1\})$ and for $V = \{x^0, x^1\}$ has the substitution polynomial

$$x^0 \begin{pmatrix} 0 & 1 & 0 & 0 \\ 0 & 0 & 0 & 1 \\ 0 & 0 & 1 & 0 \\ 0 & 0 & 0 & 1 \end{pmatrix} \oplus x^1 \begin{pmatrix} 1 & 0 & 0 & 0 \\ 0 & 0 & 1 & 0 \\ 0 & 0 & 1 & 0 \\ 0 & 0 & 0 & 1 \end{pmatrix}.$$

The substitution polynomial will be more relevant in the following sections.

2.3   Notes and comments

The question on the existence of expanding endomorphisms of a finitely generated group is closely related to the existence of expanding maps on certain Riemannian manifolds, see [49], [130], [154], [166]. It is also very much related to the work of Gromov on the growth properties of finitely generated groups, see [88].

Residue sets and complete digit sets are usually studied within the framework of representations of the integers or natural numbers, see e.g., [7], [9], [61], [79], [82], [103], [107], [152], and the references given there.

The Thue–Morse sequence appears in [158] and later independently in [131]. It is one of the best studied automatic sequences. Different facets of this sequence are discussed in: [3], [5], [13], [26], [27], [41], [51], [52], [62], [68], [70], [71], [75], [76], [81], [86], [104], [112], [113], [122], [139], [146], [150], [161], [162], [167], [168], to name but a few.

The paperfolding sequence is also a prominent example of an automatic sequence. In [15], [19], [37], [65], [66], [67], [108], [110], [125], [126], [129], [132], [144], [145], [146], [157] the reader will find detailed discussions about properties of the paperfolding sequence as well as generalizations of the paperfolding sequence.

The Baum–Sweet sequence defined in [32] seems to be less interesting.

In [11], one finds a discussion of the binomial coefficients modulo a natural number.

We conclude with some general remarks on substitutions. Substitutions provide a basic tool for the constructions of fractal sets, see, e.g.,[135], [138] and the literature cited there. Substitutions play also a prominent role in the theory of formal languages and in the theory of combinatorics on words. In this context substitutions are special morphisms of a free monoid, see, e.g., [60], [115] for more details.

Chapter 3

Automaticity

In this chapter, we introduce the concept of automatic sequences over Γ. We begin with the definition of automaticity, to be precise, (V_c, H)-automaticity of a sequence $\underline{f} \in \Sigma(\Gamma, \mathcal{A})$, where V_c is a complete digit set for H. Roughly speaking, a (V_c, H)-automatic sequence is generated by a finite (V_c, H)-automaton. We demonstrate that the set of sequences which are generated by a (V_c, H)-automaton coincides with the set of sequences with a finite (V_c, H)-kernel.

Closely related to a finite (V_c, H)-automaton is a directed labeled graph, the transition graph of the automaton. We shall show that the finite (V_c, H)-kernel graph of a sequence provides a transition graph of a (V_c, H)-automaton that generates the sequence. Furthermore, this automaton is minimal within a certain class of automata. Finally we develop an algorithm that allows us to construct a minimal (w.r.t. the number of states) (V_c, H)-automaton for a given (V_c, H)-automatic sequence.

At the end of the first part, we also indicate how to introduce a notion of (V, H)-automaticity if V is not a complete digit set for H.

The second part of this chapter is devoted to investigate properties of (V_c, H)-automatic sequences. The main result is that the automaticity of a sequence does not depend on the choice of the residue set V. In other words, if $\underline{f}$ has a finite (V, H)-kernel, then $\underline{f}$ has a finite (W, H)-kernel for all residue sets W of H. It is therefore justified to speak of an H-automatic sequence.

3.1   Automatic sequences

Again, we consider the Thue Morse sequence $(t_n)_{n \geq 0}$. As we have seen in the introduction, there exists a graph associated with the Thue Morse sequence, cf. Figure 1. This graph allows us to compute the value t_n using the binary expansion of n.

In Chapter 2 we introduced the kernel graph. In this section, we show that the kernel graph of a sequence $\underline{f}$ generated by a (V, H)-substitution allows the computation of $\underline{f}(\gamma)$ from a knowledge of the $\zeta_{V,H}$-image of the orbit of γ under iteration of $\kappa_{V,H}$.

We consider an expanding endomorphism $H : \Gamma \to \Gamma$ (w.r.t. a fixed norm $\| \; \|$) together with a complete digit set V_c, cf. Definition 2.2.5. We begin with the definition of a (V_c, H)-automatic sequence.

Definition 3.1.1. The sequence $\underline{f} \in \Sigma(\Gamma, \mathcal{A})$ is called (V_c, H)-*automatic* if there exist a finite set $\overline{\mathcal{B}} = \mathcal{B} \cup \{\varnothing\}$ and maps $\alpha_v : \overline{\mathcal{B}} \to \overline{\mathcal{B}}$, with $v \in V_c$, $\omega : \overline{\mathcal{B}} \to \mathcal{A}$ such that:

1. $\omega(\varnothing) = \underline{f}(e)$.

2. For $\gamma \in \Gamma \setminus \{e\}$ let $\gamma = v_{i_0} H(v_{i_1}) \dots H^{\circ n}(v_{i_n})$ be the digit representation, i.e., if $v_{i_n} \neq e$, then

$$\omega(\alpha_{v_{i_n}} \circ \cdots \circ \alpha_{v_{i_0}}(\varnothing)) = \underline{f}(\gamma)$$

   holds for all $\gamma \neq e$.

Remarks.

1. If it is clear from the context what V_c and H are, we simply speak of an automatic sequence $\underline{f}$.

2. If the sequence $\underline{f} \in \Sigma(\Gamma, \mathcal{A})$ is automatic, then the quintuple $\mathrm{aut}(\underline{f}) = (\overline{\mathcal{B}}, \varnothing, \mathcal{A}, \omega, \{\alpha_v \mid v \in V_c\})$ is called a (V_c, H)-*automaton* for $\underline{f}$. The set $\overline{\mathcal{B}}$ is called the *state alphabet*, or simply the *states*, the special element $\varnothing$ is called the *initial state*, the set $\mathcal{A}$ is the output alphabet, the map $\omega : \overline{\mathcal{B}} \to \mathcal{A}$ is the *output map*, and the maps $\{\alpha_v \mid v \in V\}$ are called *transition functions*. We often say that the automaton $\mathrm{aut}(\underline{f})$ generates the sequence $\underline{f}$.

   The reader should be aware of the fact that there are several different finite automata generating a sequence $\underline{f}$. Despite this fact, the notion $\mathrm{aut}(\underline{f})$ simply means that we have chosen one automaton generating $\underline{f}$.

3. Due to Corollary 2.2.6, any $\gamma \in \Gamma$ has a unique digit representation.

4. If V is any residue set for H, then Γ_e is a (V, H)-substitution-invariant subset of Γ and we can speak of the automaticity of elements in $\Sigma(\Gamma_e, \mathcal{A})$.

   The next examples provide some insight how to define a finite automaton which generates a given sequence.

Examples.

1. Consider the Thue–Morse sequence $\underline{t}$ as an element of $\Sigma(\mathbb{N}, \{0, 1\})$. Then $\underline{t}$ is (V, H)-automatic, where $H(x^j) = x^{2j}$ and $V = \{x^0, x^1\}$. The state alphabet $\overline{\mathcal{B}}$ is given by the set $\overline{\mathcal{B}} = \{\varnothing, 1\}$, the output alphabet is $\mathcal{A} = \{0, 1\}$, the output function $\omega : \overline{\mathcal{B}} \to A$ is defined by $\omega(\varnothing) = 0$ and $\omega(1) = 1$, the transition functions α_{x^0} and α_{x^1} are given by

$$\alpha_{x^0}(\varnothing) = \varnothing, \quad \alpha_{x^0}(1) = 1,$$
$$\alpha_{x^1}(\varnothing) = 1, \quad \alpha_{x^0}(1) = \varnothing.$$

   Thus the Thue–Morse sequence viewed as an element of $\Sigma(\mathbb{N}, \{0, 1\})$ is (V, H)-automatic.

2. Let $ps \in \Sigma(\mathbb{N}, \{0, 1\}$ denote the paperfolding sequence. Then ps is (V, H)-automatic, where $H(x^j) = x^{2j}$ and $V = \{x^0, x^1\}$. The state alphabet $\overline{\mathcal{B}}$ is given by the set $\overline{\mathcal{B}} = \{\varnothing, 1, 2, 3\}$, the output alphabet is $\mathcal{A} = \{0, 1\}$, the output function $\omega : \overline{\mathcal{B}} \to \mathcal{A}$ is defined by $\omega(\varnothing) = 1$, $\omega(1) = 1$, $\omega(2) = 0$, and $\omega(3) = 1$.

The transition functions α_{x^0} and α_{x^1} are given by

$$\alpha_{x^0}(\varnothing) = 1, \quad \alpha_{x^0}(1) = 3, \quad \alpha_{x^0}(2) = 2, \quad \alpha_{x^0}(3) = 3,$$
$$\alpha_{x^1}(\varnothing) = \varnothing, \quad \alpha_{x^0}(1) = 2, \quad \alpha_{x^1}(2) = 2, \quad \alpha_{x^0}(3) = 3.$$

3. The sequence $\underline{f}_0 \in \Sigma(\mathbb{N}, \{0, 1, 2\})$, see Example 4, p. 37, is (V_c, H)-automatic, where $H(x^j) = x^{3j}$ and $V_c = \{x^{-1}, x^0, x^1\}$. The state alphabet $\overline{\mathcal{B}}$ is the set $\{\varnothing, 1, 2\}$, the output alphabet $\mathcal{A}$ is $\{0, 1, 2\}$, the output function $\omega : \overline{\mathcal{B}} \to \mathcal{A}$ is defined by $\omega(\varnothing) = 0$, $\omega(1) = 1$, and $\omega(2) = 2$. The transition functions are given by

$$\alpha_{x^{-1}}(\varnothing) = 1, \quad \alpha_{x^{-1}}(1) = 2, \quad \alpha_{x^{-1}}(2) = 2,$$
$$\alpha_{x^0}(\varnothing) = \varnothing, \quad \alpha_{x^0}(1) = 1, \quad \alpha_{x^0}(2) = 2,$$
$$\alpha_{x^1}(\varnothing) = 2, \quad \alpha_{x^1}(1) = \varnothing, \quad \alpha_{x^1}(2) = 2.$$

If we consider the sequence $\underline{cs} = \theta(\underline{f}_0) \in \Sigma(\mathbb{N}, \{0, 1\})$ defined in Example 4, then the notion of (V_c, H)-automaticity makes no sense, since the map κ_{H, V_c}, where H and V_c are as above, is not defined as a map from $\mathbb{N}$ to $\mathbb{N}$.

If we choose $V = \{x^0, x^1, x^2\}$ instead of V_c, then $\underline{cs} \in \Sigma(\mathbb{N}, \{0, 1, 2\})$ is indeed (V, H)-automatic.

Definition 3.1.2. Let $f \in \Sigma(\Gamma, \mathcal{A})$ be a (V_c, H)-automatic sequence and let $\text{aut}(\underline{f}) = (\overline{\mathcal{B}}, \varnothing, \mathcal{A}, \omega, \{\alpha_v \mid v \in V_c)$ be a generating automaton. Then the finite, labeled, directed graph $G = G(\text{aut}(\underline{f})) = (\overline{\mathcal{B}}, \varnothing, V_c, K)$ with basepoint $\varnothing$ where the set of edges is given by

$$K = \left\{ (b_1 \xrightarrow{v} b_2) \mid b_1, b_2 \in \overline{\mathcal{B}} \text{ such that there exists } \alpha_v \text{ with } \alpha_v(b_1) = b_2 \right\},$$

is called the *transition graph* of $\text{aut}(\underline{f})$.

The output function $\omega : \overline{\mathcal{B}} \to \mathcal{A}$ can be considered as a function on the vertices of the graph $G(\text{aut}(\underline{f}))$. The automaton $\text{aut}(\underline{f})$ "produces" the sequence $\underline{f}$ in the following way. Let $\gamma \in \Gamma$, $\gamma \neq e$, then γ has a unique finite representation

$$\gamma = v_0 H(v_1) H^{\circ 2}(v_2) \ldots H^{\circ k}(v_k), \cdot$$

with $v_j \in V_c$ for $j = 1, \ldots, k$ and $v_k \neq e$. The representation of γ defines a path in the graph G. The path starts at the basepoint $\varnothing$ and follows the edges labeled

$v_0, v_1, \ldots, v_k$. The value of ω at the terminal vertex of the path is equal to $\underline{f}(\gamma)$.
If $\gamma = e$, then $\omega(\varnothing) = \underline{f}(e)$. We often speak simply of the transition graph of a
sequence. Figure 3.1 shows a transition graph for the Thue–Morse sequence with
output map $\omega(\varnothing) = 0$, $\omega(I) = 1$. For the sake of simplicity the arrows are labeled 0
and 1 rather then x^0, x^1.

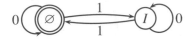

Figure 3.1. Transition graph for the Thue–Morse sequence.

If $(\overline{\mathcal{B}}_1, \varnothing_1, \mathcal{A}_1, \omega_1, \{\alpha_v \mid v \in V_c\})$ and $(\overline{\mathcal{B}}_2, \varnothing_2, \mathcal{A}_2, \omega_2, \{\beta_v \mid v \in V_c\})$ are finite
(V_c, H)-automata, then we say that they are isomorphic if their transition graphs
are isomorphic as directed, labeled graphs with basepoint, cf. the definition given on
page 49.

As a next step, we study the connection of sequences generated by a substitution
and automatic sequences. Let $\underline{f}$ be (V_c, H)-automatic. Let $\gamma \in \Gamma$, $\gamma \neq e$ and
$\gamma = v_{i_0} H(v_1) \ldots H^{\circ n}(v_{i_n})$ be its unique representation, in particular $v_{i_n} \neq e$, then
the map $\alpha_\gamma : \overline{\mathcal{B}} \to \overline{\mathcal{B}}$ is defined by

$$\alpha_\gamma = \alpha_{v_{i_n}} \circ \cdots \circ \alpha_{v_{i_0}}.$$

If κ and ζ denote the image-part-map and remainder-map of H w.r.t. V_c, respectively,
then we have $\alpha_\gamma = \alpha_{\kappa(\gamma)} \circ \alpha_{\zeta(\gamma)}$ for all $\gamma \neq e$.

If $\underline{f}$ is (V_c, H)-automatic and if $\mathrm{aut}(\underline{f}) = (\mathcal{B}, \varnothing, \mathcal{A}, \omega, \{\alpha_v \mid v \in V_c\})$ is a
generating automaton for $\underline{f}$, then the sequence $\underline{F}_\varnothing$ defined by

$$\underline{F}_\varnothing = \varnothing e \oplus \bigoplus_{\gamma \in \Gamma, \gamma \neq e} \alpha_\gamma(\varnothing)\gamma$$

is an element of $\Sigma(\Gamma, \overline{\mathcal{B}})$ and (V_c, H)-automatic. It is generated by the (V_c, H)-
automaton $\mathrm{aut}(\underline{F}_\varnothing) = (\overline{\mathcal{B}}, \varnothing, \overline{\mathcal{B}}, \mathrm{id}, \{\alpha_v \mid v \in V_c\})$. Furthermore, the output function
ω of $\mathrm{aut}(\underline{f})$ satisfies $\hat{\omega}(\underline{F}_\varnothing) = \underline{f}$.

By analogy, we define for any $b \in \overline{\mathcal{B}}$ the sequence $\underline{F}_b$ by

$$\underline{F}_b = b e \oplus \bigoplus_{\gamma \in \Gamma, \gamma \neq e} \alpha_\gamma(b)\gamma. \tag{3.1}$$

Definition 3.1.3. Let $\mathcal{A}$ be a finite set and $\Sigma(\Gamma, \mathcal{A})$ the set of sequences. The map
$L : \Sigma(\Gamma, \mathcal{A}) :\to \Sigma(\Gamma, \overline{\mathcal{A}})$ defined by

$$L(\underline{f}) = \varnothing \oplus \bigoplus_{\gamma \neq e} f_\gamma \, \gamma$$

is called *L-map*.

Note that any sequence $\underline{f} \in \Sigma(\Gamma, \mathcal{A})$ can be written as $\underline{f} = f_e\, e \oplus L(\underline{f})$, where the summands have to be considered as elements in $\Sigma(\Gamma, \overline{\mathcal{A}})$.

The following lemma is immediate, it relates the L-map and the decimations.

Lemma 3.1.4. *If $\underline{g} \in \Sigma(\Gamma, \mathcal{A})$ and $v \in V_c$, $v \neq e$, then*

$$\partial_v^H(L(\underline{g})) = \begin{cases} \partial_v^H(\underline{g}) & \text{if } v \neq e \\[2mm] L(\partial_e^H(\underline{g})) & \text{if } v = e. \end{cases}$$

In other words, the L-map commutes with ∂_e^H, and ∂_v^H, $v \neq e$, is a left inverse of the L-map.

Let $\underline{f}$ be generated by $(\overline{\mathcal{B}}, \varnothing, \mathcal{A}, \omega, \{\alpha_v \mid v \in V\})$. For $a, b \in \overline{\mathcal{B}}$ we define the sequences $\underline{G}_{a,b} \in \Sigma(\Gamma, \overline{\mathcal{B}})$ by

$$\underline{G}_{a,b} = ae \oplus L(\underline{F}_b), \tag{3.2}$$

where $\underline{F}_b$ is defined as in Equation (3.1). The next lemma relates the maps α_v of $(\overline{\mathcal{B}}, \varnothing, \mathcal{A}, \omega, \{\alpha_v \mid v \in V_c\})$ to the decimations ∂_v^H of $\underline{G}_{a,b}$.

Lemma 3.1.5. *Let $(\overline{\mathcal{B}}, \varnothing, \overline{\mathcal{B}}, \text{id}, \{\alpha_v \mid v \in V_c\})$ be a (V_c, H)-automaton. If $\underline{G}_{a,b}$ is defined as in Equation (3.2), then*

$$\partial_v^H(\underline{G}_{a,b}) = \begin{cases} \underline{G}_{\alpha_v(b),\alpha_v(b)} & \text{if } v \neq e \\ \underline{G}_{a,\alpha_e(b)} & \text{if } v = e. \end{cases}$$

Proof. By the definition of $\underline{G}_{a,b}$, one has

$$\underline{G}_{a,b} = ae \uplus \bigoplus_{\gamma \neq e} \alpha_\gamma(b)\, \gamma,$$

and therefore

$$\underline{G}_{a,b} = ae \oplus \bigoplus_{\substack{v \in V_c, \gamma \in \Gamma, \\ vH(\Gamma) \neq e}} \alpha_{vH(\gamma)(b)}\, vH(\gamma).$$

For $w \in V_c$ with $w \neq e$ this gives

$$\partial_w^H(\underline{G}_{a,b}) = \bigoplus_{\gamma \in \Gamma} \alpha_{wH(\Gamma)}(b)\, \gamma = \bigoplus_{\gamma \in \Gamma} \alpha_\gamma(\alpha_w(b))\, \gamma = \underline{G}_{\alpha_w(b),\alpha_w(b)}.$$

This proves the first part of the assertion. The ∂_e^H-decimation of $\underline{G}_{a,b}$ is given by

$$\partial_e^H(\underline{G}_{a,b}) = ae \oplus \bigoplus_{\gamma \in \Gamma, \gamma \neq e} \alpha_{H(\gamma)}(b)\, \gamma = ae \oplus \bigoplus_{\gamma \in \Gamma, \gamma \neq e} \alpha_\gamma(\alpha_e(b))\, \gamma = \underline{G}_{a,\alpha_e(b)}.$$

This proves the second assertion. $\qquad\qquad\qquad\qquad\qquad\qquad\qquad \square$

Theorem 3.1.6. $\underline{f} \in \Sigma(\Gamma, \mathcal{A})$ *is* (V_c, H)-*automatic if and only if* $\ker_{V_c, H}(\underline{f})$ *is finite.*

Proof. Let us suppose that $\underline{f}$ is automatic, i.e., there exists a (V_c, H)-automaton $(\overline{\mathcal{B}}, \varnothing, \mathcal{A}, \omega, \{\alpha_v \mid v \in V_c\})$. Then one has $\underline{f} = \hat{\omega}(\underline{F}_\varnothing) = \hat{\omega}(\underline{G}_{\varnothing,\varnothing})$. This yields for $v \subset V_c$

$$\partial_v(\underline{f}) = \hat{\omega}(\partial_v(\underline{G}_{\varnothing,\varnothing})).$$

By Lemma 3.1.5, it follows that every kernel element $\underline{h}$ of $\underline{f}$ can be written as $\hat{\omega}(\underline{G}_{a,b})$, where $a, b \in \overline{\mathcal{B}}$ are determined by $\underline{h}$. Therefore the (V_c, H)-kernel of $\underline{f}$ is finite.

Suppose that $\underline{f}$ possesses a finite (V_c, H)-kernel. We construct a finite automaton generating $\underline{f}$. To this end, let $\overline{\mathcal{B}} = \ker_{V_c, H}(\underline{f}) = \{\underline{f}_1 = \underline{f}, \ldots, \underline{f}_N\}$ be the state alphabet with distinguished element $\varnothing = \underline{f}$. The output alphabet is the set $\mathcal{A}$, the output function $\omega : \overline{\mathcal{B}} \to \mathcal{A}$ is defined by $\omega(\underline{h}) = \underline{h}(e)$. The transition functions $\alpha_v : \overline{\mathcal{B}} \to \overline{\mathcal{B}}$ are defined by $\alpha_v(\underline{h}) = \partial_v^H(\underline{h})$. Thus, we have $\omega(\varnothing) = \omega(\underline{f}) = \underline{f}(e)$ and for $\gamma \in \Gamma \setminus \{e\}$ with digit representation $\gamma = v_{i_0} H(v_{i_1}) \ldots H^{\circ n}(v_{i_n})$, we have

$$\omega(\alpha_{v_{i_n}} \circ \cdots \circ \alpha_{v_{i_0}}(\varnothing)) = [\partial_{v_n}^H \circ \cdots \circ \partial_{v_{i_0}}^H(\underline{f})](e) = \underline{f}(v_{i_0} H(v_{i_1}) \ldots H^{\circ n}(v_{i_n})).$$

Therefore $\underline{f}$ is (V_c, H)-automatic.   □

The proof of Theorem 3.1.6 shows that the kernel graph of a sequence can be interpreted as a transition graph of an automaton. Therefore the kernel graph provides a 'natural' finite automaton generating the sequence $\underline{f}$. The output function is then defined by $\theta : \ker_{V_c, H}(\underline{f}) \to \mathcal{A}$ with $\theta(\underline{h}) = \underline{h}(e)$.

Note further that the first part of the proof of Theorem 3.1.6 also provides a way to compute the kernel graph of the sequence $\underline{G}_{\varnothing,\varnothing} = \underline{F}_\varnothing$.

Example. Let $\Gamma = \mathbb{N}$ and $H(x^j) = x^{2j}$ with $V_c = \{x^0, x^1\}$ as complete digit set. Figure 3.2 shows a transition graph of a finite (V_c, H)-automaton. The first terms of

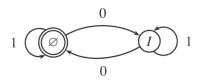

Figure 3.2. Transition graph of a (V_c, H)-automaton.

the sequence $\underline{F}_\varnothing$ are then given by

$$\varnothing\varnothing I\varnothing\varnothing II\varnothing\ldots.$$

Applying Lemma 3.1.5, we can compute the (V_c, H)-kernel graph of $\underline{F}_\varnothing$. It is shown in Figure 3.3.

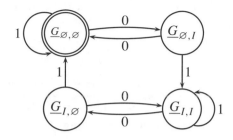

Figure 3.3. Kernel graph of $\underline{F}_{\varnothing} = \underline{G}_{\varnothing,\varnothing}$ generated by the automaton in Figure 3.2.

We now investigate the relation between $|\ker_{V,h}(\underline{f})|$ and the cardinality of the state set $\overline{\mathcal{B}}$ of $\text{aut}(\underline{F})$.

Definition 3.1.7. Let $(\overline{\mathcal{B}}, \varnothing, \mathcal{A}, \omega, \{\alpha_v \mid v \in V_c\})$ be a finite (V_c, H)-automaton. Let $a, b \in \overline{\mathcal{B}}$. The state a is *accessible* from b if $a = b$, or if there exist $n \in \mathbb{N}$ and $v_{i_0}, \ldots, v_{i_n} \in V_c$ such that $a = \alpha_{v_{i_0}} \circ \cdots \circ \alpha_{v_{i_n}}(b)$. The set

$$\text{Acc}(b, \overline{\mathcal{B}}, \{\alpha_v \mid v \in V_c\}) = \{a \in \overline{\mathcal{B}} \mid a \text{ is accessible from } b\}$$

is called the *set of accessible states* (from b w.r.t. the automaton).

Remarks.
1. If the automaton is clear from the context, we simply write $\text{Acc}(b)$ for the set of accessible states. In particular, $\text{Acc}(\underline{f})$ is the set of accessible elements of $\ker_{V,H}(\underline{f})$ (w.r.t. the v-decimations).

2. The set of accessible states $\text{Acc}(b)$ is the smallest subset of $\overline{\mathcal{B}}$ which is invariant under all maps α_v, $v \in V_c$, and contains b.

As a consequence of Lemma 3.1.5 we obtain the following estimate on the number of kernel elements.

Proposition 3.1.8. *Let* $\underline{f} \in \Sigma(\Gamma, \mathcal{A})$ *be generated by the finite automaton* $(\overline{\mathcal{B}}, \varnothing, \mathcal{A}, \omega, \{\alpha_v \mid v \in V_c\})$, *then*

$$\left| \ker_{V_c,H}(\underline{f}) \right| \leq |\text{Acc}(\varnothing)|^2.$$

Examples.
1. Let $\underline{f} \in \Sigma(\mathbb{N}, \{0, 1\})$ be defined by

$$\underline{f} = \bigoplus_{n \geq 0} f_n x^n,$$

where $f_n = 1$ if and only if $n = 2^k - 1$ for $k \in \mathbb{N}$, otherwise $f_n = 0$. In a more compact way we write

$$\underline{f} = \bigoplus_{n \geq 0} x^{2^n - 1},$$

where we agree that all other values of the sequence $\underline{f}$ are equal to zero.

Then $\underline{f}$ is (V_c, H)-automatic, where $V_c = \{x^0, x^1\}$ and $H(x^j) = x^{2j}$. The transition graph of the automaton generating $\underline{f}$ is shown in Figure 3.4, the

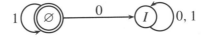

Figure 3.4. Transition graph of the automaton for $\underline{f} = \bigoplus_{n=0}^{\infty} x^{2^n - 1}$.

output function is $\omega(\varnothing) = 1$ and $\omega(I) = 0$. We then compute

$$\partial_{x^0}^H(\underline{f}) = \partial_{x^0}^H(1\,x^0 \oplus L(\underline{F}_1)) = 1\,x^0 \oplus L(\underline{F}_{\alpha_{x^0}(1)}) = 1\,x^0 \oplus L(\underline{F}_0) = \underline{g}$$

$$\partial_{x^1}^H(\underline{f}) = \partial_{x^1}^H(1\,x^0 \oplus L(\underline{F}_1)) = \alpha_{x^1}(1)\,x^0 \oplus L(\underline{f}_{\alpha_{x^1}(1)}) = 1\,x^0 \oplus L(\underline{F}_1) = \underline{f}$$

$$\partial_{x^0}^H(1\,x^0 \oplus L(\underline{F}_0)) = 1\,x^0 \oplus L(\underline{f}_{\alpha_{x^0}(0)}) = 1\,x^0 \oplus L(\underline{F}_0) = \underline{g}$$

$$\partial_{x^1}^H(1\,x^0 \oplus L(\underline{F}_0)) = 0\,x^0 \oplus L(\underline{F}_{\alpha_{x^1}(0)}) = 0\,x^0 \oplus L(\underline{F}_0) = \underline{h},$$

and $\partial_{x^0}^H(\underline{h}) = \partial_{x^1}^H(\underline{h}) = \underline{h}$. The kernel-graph is shown in Figure 3.5.

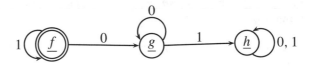

Figure 3.5. Kernel graph of $\underline{f}$ generated by the automaton in Figure 3.4.

2. Let $\overline{\mathcal{B}} = \{\varnothing, A, B, C\}$ and let $(\overline{\mathcal{B}}, A, \mathrm{id}, \mathcal{A}, \{\alpha_v \mid v \in V_c\})$ be a finite automaton with H and V_c as above. The transition functions are given by the graph in Figure 3.6.

The kernel graph of the sequence $\underline{F}_\varnothing$ is shown in Figure 3.7.

Note that there are 7 different kernel elements while the generating automaton has only 4 different states.

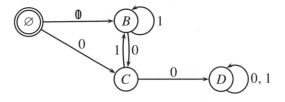

Figure 3.6. Finite automaton for the second example.

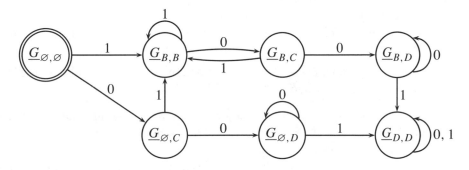

Figure 3.7. Kernel graph for the sequence $\underline{F}_\varnothing$ of the second example.

As we have seen already, the square of the cardinality of the set

$$\mathrm{Acc}(\varnothing, \overline{\mathscr{B}}, \{\alpha_v \mid v \in V_c\}))$$

is an upper bound for the number of kernel elements of $\underline{F}_\varnothing$.

The range, $\mathscr{R}(\underline{F}_\varnothing)$, see Definition 2.2.21, of the sequence $\underline{F}_\varnothing$ is a subset of the accessible states. Using the range and the dynamics of α_e as a map from $\overline{\mathscr{B}}$ to $\overline{\mathscr{B}}$ we can obtain a better estimate on the number of kernel elements of $\underline{F}_\varnothing$.

Lemma 3.1.9. *Let* $f \in \Sigma(\Gamma, \mathscr{A})$ *be generated by the automaton* $\mathrm{aut}(\underline{f}) = (\overline{\mathscr{B}}, \varnothing, \mathscr{A}, \omega, \{\alpha_v \mid v \in V_c\})$, *then*

$$\left| \ker_{V_c, H}(\underline{F}_\varnothing) \right| \leq \sum_{b \in \mathscr{R}(\underline{F}_\varnothing)} |\mathcal{O}_e(b)|,$$

where $\mathcal{O}_e(b) = \{\alpha_e^{\circ n}(b) \mid n \in \mathbb{N}\}$ *denotes the orbit of* b *under iteration of* α_e, $\mathscr{R}(\underline{F}_\varnothing)$ *denotes the range of* $\underline{F}_\varnothing$, *and* $\underline{F}_\varnothing$ *is defined as in Equation* (3.1).

Proof. The set K defined by

$$K = \{\underline{G}_{a,b} = b\,e \oplus L(\underline{F}_{\alpha_e^{\circ k}(b)}) \mid b \in \mathscr{R}(\underline{F}_\varnothing), \, k \in \mathbb{N}\}$$

is a finite set, it contains $\underline{G}_{\varnothing,\varnothing} = \underline{F}_\varnothing$ and is, due to Lemma 3.1.5, invariant under decimations. $\qquad\square$

Remarks.

1. Note that Lemma 3.1.9 yields the estimate

$$\left|\ker_{V_c,H}(\underline{f})\right| \le \sum_{b \subset \mathcal{R}(\underline{F}_\varnothing)} |\mathcal{O}_e(b)|,$$

since $\hat{\omega}(\underline{F}_\varnothing) = \underline{f}$. This is the best possible estimate.

2. Example 2, see also Figure 3.6, from above shows that the estimate stated in Lemma 3.1.9 is sharp. The range of $\underline{F}_\varnothing$ is the set $\{\varnothing, B, D\}$ and the orbits under iteration of α_e are given by $\varnothing \mapsto C \mapsto D \mapsto D$, $B \mapsto C \mapsto D \mapsto D$, and $D \mapsto D$. This gives 7 as an upper bound for the number of kernel elements. And indeed, the kernel has cardinalty 7, see also Figure 3.7.

   Let $\omega : \{\varnothing, B, C, D\} \to \{0, 1\}$ be defined by $\omega(z) = 1$ if and only if $z = D$. The sequence $\underline{f} = \hat{\omega}(\underline{F}_\varnothing)$ has also 7 kernel elements.

As we have noted already there exist several different finite automata which generate a given sequence $\underline{f}$. It is therefore natural to ask for the minimal finite automaton that generates the sequence $\underline{f}$. An automaton $(\mathcal{B}, \varnothing, \mathcal{A}, \omega, \{\alpha_v \mid v \in V_c\})$ that generates $\underline{f}$ is called minimal if the number of states is minimal. It is called an $\underline{f}$-*automaton*.

In the following we discuss properties of minimal automata. We show that the kernel graph provides a minimal automaton in a certain subclass of finite automata. Then we present a method how to obtain a finite automaton which is minimal. We begin with an almost trivial estimate on the minimal number of states.

Lemma 3.1.10. *Let $\underline{f} \in \Sigma(\Gamma, \mathcal{A})$ be (V_c, H)-automatic. If $\mathrm{aut}(\underline{f})$ is a minimal (V_c, H)-automaton, then*

$$\left|\mathcal{R}(\underline{f})\right| \le |\mathcal{B}| \le \left|\ker_{V_c,H}(\underline{f})\right|.$$

Proof. If $\omega : \overline{\mathcal{B}} \to \mathcal{A}$ denotes the output map of a minimal automaton, then $\mathcal{R}(\underline{f}) \subset \omega(\overline{\mathcal{B}})$, i.e., $\overline{\mathcal{B}}$ contains at least $\left|\mathcal{R}(\underline{f})\right|$ different states.

As we have already seen, the (V_c, H)-kernel graph of $\underline{f}$ provides a finite automaton that generates $\underline{f}$. Thus the number of states of a minimal automaton is at most $\left|\ker_{V_c,H}(\underline{f})\right|$.   □

The next lemma collects some necessary conditions for an automaton $\mathrm{aut}(\underline{f})$ to be minimal.

Lemma 3.1.11. *If $\mathrm{aut}(\underline{f}) = (\overline{\mathcal{B}}, \varnothing, \mathcal{A}, \omega, \{\alpha_v \mid v \in V_c\})$ is a minimal (V_c, H)-automaton for $\underline{f}$, then $\mathrm{aut}(\underline{f})$ has the following properties:*

1. *The set of accessible states $\mathrm{Acc}(\varnothing, \overline{\mathcal{B}}, \{\alpha_v \mid v \in V_c\}))$ is equal to the set of states, i.e., $\overline{\mathcal{B}} = \mathrm{Acc}(\varnothing)$.*

2. *If $a, b \in \overline{\mathscr{B}}$ such that $a \neq b$, then $\hat{\omega}(\underline{F}_a) \neq \hat{\omega}(\underline{F}_b)$ (see Equation (3.1) for the definition of $\underline{F}_a$).*

Proof. 1. If there exists $b \in \overline{\mathscr{B}}$ such that $b \notin \mathrm{Acc}(\varnothing)$, then the set

$$\overline{\mathscr{B}}' = \mathrm{Acc}(\varnothing)$$

is a proper subset of $\overline{\mathscr{B}}$ and the automaton $(\overline{\mathscr{B}}', \varnothing, \omega|_{\overline{\mathscr{B}}'}, \{\alpha_v|_{\overline{\mathscr{B}}'} \mid v \in V_c\})$ is well defined and generates $\underline{f}$. In fact, it even generates $\underline{F}_\varnothing$. Since the cardinality of $\overline{\mathscr{B}}'$ is smaller than the cardinality of $\overline{\mathscr{B}}$, the automaton with states $\overline{\mathscr{B}}$ is not minimal which is a contradiction.

2. Assume that there exist $a, b \in \overline{\mathscr{B}}$ such that $a \neq b$ and $\hat{\omega}(\underline{F}_a) = \hat{\omega}(\underline{F}_b)$. Then define $\overline{\mathscr{B}}' = \overline{\mathscr{B}} \setminus \{a\}$ and new transition functions $\alpha_v' : \overline{\mathscr{B}}' \to \overline{\mathscr{B}}'$ by

$$\alpha_v'(x) = \begin{cases} \alpha_v(x) & \text{if } \alpha_v(x) \neq a \\ b & \text{if } \alpha_v(x) = a, \end{cases}$$

and $(\overline{\mathscr{B}}', \varnothing, \omega|_{\overline{\mathscr{B}}'}, \{\alpha_v' \mid v \in V_c\})$ is a finite automaton that generates the sequence $\underline{F}_\varnothing$. This is a contradiction to the minimality of $\mathrm{aut}(\underline{f})$. □

These necessary conditions for an automaton to be minimal are far from being sufficient. The first condition states that the automaton has no superfluous states. The second condition means that all states which are in a certain sense necessary appear only once.

Remarks.

1. Given an automaton which generates a sequence $\underline{f}$ and which does not satisfy condition 1 or condition 2 of Lemma 3.1.11, then it is possible to derive a smaller automaton by applying the reduction process outlined in the proof.

2. The automaton defined by the (V_c, H)-kernel graph of a sequence $\underline{f}$ is an example of an automaton satisfying both minimality conditions.

If $\mathrm{aut}(\underline{f})$ is an automaton which satisfies the above minimality conditions, then it is not easy to further reduce the number of states. The next theorem shows that the automaton which is defined by the kernel graph is minimal under the class of output-consistent automata.

Definition 3.1.12. Let $\mathrm{aut}(\underline{f}) = (\overline{\mathscr{B}}, \varnothing, \mathscr{A}, \omega, \{\alpha_v \mid v \in V_c\})$ be a (V_c, H)-automaton for $\underline{f}$. Then $\mathrm{aut}(\underline{f})$ is called *output-consistent* if for all $a, b \in \overline{\mathscr{B}}$ with $\alpha_e(a) = b$ we have that $\omega(a) = \omega(b)$.

Remark. The automaton defined by the (V_c, H)-kernel graph of $\underline{f}$ with output function $\theta : \ker_{V_c, H}(\underline{f}) \to \mathscr{A}$ defined by $\theta(\underline{h}) = \underline{h}(e)$ is an example of an output-consistent automaton.

Among the output-consistent automata there exists a minimal automaton. The following theorem characterizes the minimal output-consistent automaton.

Theorem 3.1.13. *Let $f \in \Sigma(\Gamma, \mathcal{A})$ be a sequence such that the (V_c, H)-kernel of f is finite. Then the $\overline{(V_c, H)}$-automaton defined by the (V_c, H)-kernel graph and the output function $\theta : \ker_{V_c, H}(f) \to \mathcal{A}$, where $\theta(\underline{h}) = \underline{h}(e)$, is minimal under all output-consistent automata that generate f.*

Proof. Let us assume that $\mathrm{aut}(f) = (\overline{\mathcal{B}}, \varnothing, \mathcal{A}, \omega, \{\alpha_v \mid v \in V_c\})$ is output-consistent. Let us further assume that the transition graph of $\mathrm{aut}(f)$ satisfies the minimality condition 1. of Lemma 3.1.11, or, equivalently $\overline{\mathcal{B}} = \mathrm{Acc}(\varnothing)$. For $b \in \overline{\mathcal{B}}$ we define $\underline{F}_b$ as in Equation (3.1). Then the map $\Omega : \overline{\mathcal{B}} \to \Sigma(\Gamma, \mathcal{A})$ given by

$$\Omega(b) = \hat{\omega}(\underline{F}_b)$$

is well defined. From Lemma 3.1.5, we conclude that

$$\partial_v^H(\Omega(b)) = \Omega(\alpha_v(b))$$

holds for all $v \in V_c$, $v \neq e$. For $v = e$ Lemma 3.1.5 yields

$$\partial_e^H(\Omega(b)) = \hat{\omega}(b \oplus L(\underline{F}_{\alpha_e(b)})) = \omega(b) \oplus \hat{\omega}(L(\underline{F}_{\alpha_e(b)})).$$

Since ω is output consistent, we can replace $\omega(b)$ by $\omega(\alpha_e(b))$ which gives

$$\partial_e^H(\Omega(b)) = \omega(\alpha_e(b) \oplus \hat{\omega}(L(\underline{F}_{\alpha_e(b)}))) = \hat{\omega}(\underline{F}_{\alpha_e(b)}) = \Omega(\alpha_e(b)).$$

Thus, we have $\partial_v^H(\Omega(b)) = \Omega(\alpha_v(b))$ for all $v \in V_c$ and all $b \in \overline{\mathcal{B}}$.

To prove that the automaton defined by the (V_c, H)-kernel graph is minimal it suffices to show that for any $\underline{h} \in \ker_{V_c, H}(f)$ there exists a state $b \in \overline{\mathcal{B}}$ such that $\Omega(b) = \underline{h}$. To this end, let $\underline{h} \in \ker_{V_c, H}(f)$. If $\underline{h} = f$, then choose $b = \varnothing$. If $\underline{h} = \partial_{v_k}^H \circ \cdots \circ \partial_{v_0}^H(f)$, then by our above observation we see that

$$\underline{h} = \hat{\omega}(\partial_{v_k}^H \circ \cdots \circ \partial_{v_0}^H(\underline{F}_\varnothing)) = \partial_{v_k}^H \circ \cdots \circ \partial_{v_0}^H(\Omega(\underline{F}_\varnothing)) = \Omega(\alpha_{v_k} \circ \cdots \circ \alpha_{v_0}(\varnothing)),$$

which means that $\ker_{V_c, H}(f) \subset \Omega(\overline{\mathcal{B}})$, i.e., $|\overline{\mathcal{B}}| \geq |\ker_{V_c, H}(f)|$. $\square$

Remarks.

1. Note that the above proof also shows that the automaton defined by the kernel graph is the unique, up to isomorphism, among the minimal and output-consistent automata. This follows from

$$\partial_v(\Omega(b)) = \Omega(\alpha_v(b)).$$

2. All of the above remains true if we consider an arbitrary residue set V for H and study sequences in $\Sigma(\Gamma_e, \mathcal{A})$.

Example. Let $\Gamma = \mathbb{N}$ and $H(x^j) = x^{2j}$, $V = \{x^0, x^1\}$ and define $\underline{f} \in \Sigma(\mathbb{N}, \{0, 1\})$ by the transition graph shown in Figure 3.8, where the output function is given by $\omega(\varnothing) = 0$, $\omega(I) = 1$. Note further that the automaton is certainly minimal.

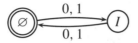

Figure 3.8. Finite automaton for the sequence $\underline{f} = \bigoplus_{n=0}^{\infty} ([\log_2(2n + 1)] \bmod 2) x^n$.

From the transition graph one easily concludes that the sequence $\underline{f} = \oplus f_n x^n = \hat{\omega}(\underline{F}_\varnothing)$ can be computed: $f_n \equiv [\log_2(2n + 1)] \bmod 2$.

Figure 3.9 shows the kernel graph of the sequence $\underline{F}_\varnothing$ which is via application of ω also the kernel graph of $\hat{\omega}(\underline{F}_\varnothing)$. It is the minimal output-consistent automaton and has as many as $|\mathrm{Acc}(\varnothing)|^2 = 4$ states.

Moreover, it can be shown that any sequence $\underline{g} \in \Sigma(\mathbb{N}, \{0, \ldots, M - 1\})$ with $n_n \equiv [\log_k(kn+1)] \bmod M$, is generated by a $(\{x^0, \ldots, x^{k-1}\}, H)$-automaton, where $H(x^j) = x^{kj}$. The minimal automaton generating the sequence has M states and the associated minimal output-consistent automaton has M^2 different states.

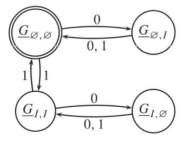

Figure 3.9. The kernel graph of $\underline{f} = \bigoplus_{n=0}^{\infty} ([\log_2(2n + 1)] \bmod 2) x^n$.

The still unsolved question is how to construct a minimal (V_c, H)-automaton that generates a given sequence $\underline{f} \in \Sigma(\Gamma, \mathcal{A})$. We tackle this problem in the following. It will turn out that the kernel graph again provides a tool to construct a minimal (V_c, H)-automaton for $\underline{f}$.

If $\underline{f}$ has a finite (V_c, H)-kernel, then the kernel graph of $\underline{f}$ is a transition graph of a finite automaton. An element $\underline{h} \in \ker_{V_c, H}(\underline{f})$ is called accessible if $\underline{h}$ is accessible from $\underline{f}$ in the sense of Definition 3.1.7, where the automaton is defined by the kernel graph. By its construction, each kernel element is accessible.

If a (V_c, H)-automaton satisfies the minimality requirements of Lemma 3.1.11, then each state is accessible. In order to distinguish between different kinds of accessibility we introduce the properly accessible states.

Definition 3.1.14. Let $(\overline{\mathcal{B}}, \varnothing, \mathcal{A}, \omega, \{\alpha_v \mid v \in V_c\})$ be a finite (V_c, H)-automaton. Let $a, b \in \overline{\mathcal{B}}$. We say that a is *properly accessible* from b (w.r.t. the automaton) if there exist $n \in \mathbb{N}$ and $v_{i_0}, \ldots, v_{i_n} \in V_c$ such that $a = \alpha_{v_{i_0}} \circ \cdots \circ \alpha_{v_{i_n}}(b)$ and $v_{i_0} \neq e$. The set

$$\text{Acc}^{\text{prop}}(b, \overline{\mathcal{B}}, \{\alpha_v \mid v \in V_c\}) = \{a \in \overline{\mathcal{B}} \mid a \text{ is properly accessible from } b\} \cup \{b\}$$

is called the *set of properly accessible states* (from b). By definition, we set $b \in \text{Acc}^{\text{prop}}(b)$.

Again we agree to write $\text{Acc}^{\text{prop}}(b)$ if all other data are clear from the context. Especially, we simply write $\text{Acc}^{\text{prop}}(\underline{f})$ for $\text{Acc}^{\text{prop}}(\underline{f}, \ker_{V_c, H}(\underline{f}), \{\partial_v^H \mid v \in V_c\})$, the properly accessible states from $\underline{f}$ w.r.t. the automaton defined by the kernel graph, if $\underline{f}$ has a finite (V_c, H)-kernel.

The L-map introduced in Definition 3.1.3 introduces an equivalent relation on $\Sigma(\Gamma, \mathcal{A})$.

Definition 3.1.15. Two sequences $\underline{g}, \underline{h} \in \Sigma(\Gamma, \mathcal{A})$ are called *L-equal* if $L(\underline{g}) = L(\underline{h})$.

It is obvious that the relation L-equal is an equivalence relation. The equivalence class of a sequence $\underline{h}$ is denoted by $[\underline{h}]_L$.

Theorem 3.1.16. *Let $\underline{f} \in \Sigma(\Gamma, \mathcal{A})$ be generated by the automaton*

$$(\overline{\mathcal{B}}, \varnothing, \mathcal{A}, \omega, \{\alpha_v \mid v \in V_c\})$$

which satisfies condition 1. *of Lemma* 3.1.11. *Then*

$$|\overline{\mathcal{B}}| \geq \left| \text{Acc}^{\text{prop}}(\underline{f}) \right| + \left| K^*(\underline{f}) \right|,$$

where $\text{Acc}^{\text{prop}}(\underline{f})$ *is the set of properly accessible* $\underline{h} \in \ker_{V_c, H}(\underline{f})$ *and* $K^*(\underline{f})$ *is the set*

$$K^*(\underline{f}) = \left\{ [\underline{h}]_L \mid \underline{h} \in \ker_{V_c, H}(\underline{f}) \text{ and } L(\underline{h}) \notin L\big(\ker_{V_c, H}(\underline{f})\big) \right\}.$$

Proof. Let $\underline{h} \in \text{Acc}^{\text{prop}}$, i.e., $\underline{h} = \partial_{v_k}^H \circ \cdots \circ \partial_{v_0}(\underline{f})$ and $v_k \neq e$. This can be written as

$$\underline{h} = \hat{\omega}(\partial_{v_k}^H \circ \cdots \circ \partial_{v_0}^H(F_\varnothing)).$$

A repeated application of Lemma 3.1.5 gives

$$\underline{h} = \hat{\omega}\big(\underline{F}_{\alpha_{v_k} \circ \cdots \circ \alpha_{v_0}(\varnothing)}\big).$$

This proves that every properly accessible kernel element requires a state in $\overline{\mathcal{B}}$.

Now let $\underline{h} \in \ker_{V_c, H}(\underline{f})$ be not properly accessible, i.e.,

$$\underline{h} = \hat{\omega}(\partial_e^H \circ \partial_{v_k}^H \circ \cdots \circ \partial_{v_0}^H (\underline{F}_\varnothing)).$$

A repeated application of Lemma 3.1.5 then gives

$$\underline{h} = \hat{\omega}\left(\underline{G}_{a, \alpha_e \circ \alpha_{v_k} \circ \cdots \circ \alpha_{v_0}(\varnothing)}\right),$$

where $a \in \overline{\mathcal{B}}$ is determined by $\underline{h}$. This shows that a non-properly accessible kernel element $\underline{h}$ defines an equivalence class $[\underline{h}]_L$. Then either there exists a $\underline{g} \in \mathrm{Acc}^{\mathrm{prop}}(\underline{f})$ such that $\underline{g} \in [\underline{h}]_L$ or no such $\underline{g}$ exists. In the second case it follows that $\underline{h} \in K^*(\underline{f})$. This shows that $\overline{\mathcal{B}}$ contains additional $|K^*(\underline{f})|$ elements. This completes the proof. $\square$

It remains to show that for an (V_c, H)-automatic sequence $\underline{f} \in \Sigma(\Gamma, \mathcal{A})$ there exists an automaton such that the state set has cardinality equal to the lower bound in Theorem 3.1.16.

Theorem 3.1.17. *Let $\underline{f} \in \Sigma(\mathcal{A}, \Gamma)$ be (V_c, H)-automatic. There exists a (V_c, H)-automaton $(\overline{\mathcal{B}}, \varnothing, \mathcal{A}, \omega, \{\alpha_v \mid v \in V\})$ that generates $\underline{f}$ and satisfies*

$$|\overline{\mathcal{B}}| = |\mathrm{Acc}^{\mathrm{prop}}(\underline{f})| + |K^*(\underline{f})|.$$

Proof. The proof is constructive. Let

$$\ker_{V_c, H}(\underline{f})/L = \{[\underline{g}]_L \mid \underline{g} \in \ker_{V_c, H}(\underline{f})\}$$

denote the kernel of $\underline{f}$ modulo L-equivalence. Then fix a map $\Psi : \ker_{V_c, H}/L \to \mathrm{Acc}^{\mathrm{prop}} \cup K^*(\underline{f})$ such that

$$\left[\Psi([\underline{h}]_L)\right]_L = [\underline{h}]_L$$

holds for all $[\underline{h}]_L \in \ker_{V_c, H}(\underline{f})/L$. Note that $\Psi([\underline{h}]_L)$ can be considered as a representative in $\mathrm{Acc}^{\mathrm{prop}} \cup K^*(\underline{f})$ of the L-equivalence class $[\underline{h}]_L$.

We set $\overline{\mathcal{B}} = \mathrm{Acc}^{\mathrm{prop}}(\underline{f}) \cup K^*(\underline{f})$, where $K^*(\underline{f})$ is defined as in Theorem 3.1.16. The special element $\varnothing$ of $\overline{\mathcal{B}}$ is $\underline{f}$.

The transition functions $\alpha_v : \overline{\mathcal{B}} \to \overline{\mathcal{B}}$, $v \in V_c$ are defined in different steps.

1. For $v \neq e$ and $\underline{h} \in \mathrm{Acc}^{\mathrm{prop}}(\underline{f})$ we define

$$\alpha_v(\underline{h}) = \partial_v^H(\underline{h}).$$

Note that $\alpha_v(\underline{h}) \in \mathrm{Acc}^{\mathrm{prop}}(\underline{f})$.

2. If $v \neq e$ and $[\underline{h}]_L \in K^*(\underline{f})$, then we set

$$\alpha_v([\underline{h}]_L) = \partial_v^H(\underline{h}).$$

Note that this definition is independent of the chosen sequence in the equivalence class. Note further that $\alpha_v([\underline{h}]_L) \in \text{Acc}^{\text{prop}}(\underline{f})$.

3. Let $v = e$ and $\underline{h} \in \text{Acc}^{\text{prop}}(\underline{f})$. Then

$$\alpha_e(\underline{h}) = \Psi\big([\partial_e^H(\underline{h})]_L\big).$$

Note that $\alpha_e(\underline{h}) = a\,e \oplus L(\partial_e^H(\underline{h}))$, where $a \in \mathcal{A}$ is determined by Ψ.

4. Let $v = e$ and $[\underline{h}]_L \in K^*(\underline{f})$. Then

$$\alpha_e([\underline{h}]_L) = \Psi\big([\partial_e^H \circ \Psi([\underline{h}]_L)]_L\big).$$

Note that for every $\underline{g} \in ([\underline{h}]_L \cap \ker_{V_c,H}(\underline{f}))$ there exists an $g_e \in \mathcal{A}$ such that

$$\partial_e^H(\underline{g}) = g_e\,e \oplus L(\alpha_e([\underline{h}]_L))$$

or, equivalently,

$$\alpha_e([\underline{h}]_L) = a\,e \oplus L(\partial_e^H(\underline{g})),$$

where $a \in \mathcal{A}$ is determined by Ψ.

Due to the above definition of the transition functions α_v, $v \in V_c$, it is easy to see that for all $\underline{g} \in \text{Acc}^{\text{prop}}(\underline{f})$, for all $n \in \mathbb{N}$ and for all $v \in V_c \setminus \{e\}$ the following equation is true:

$$\partial_v^H \circ \big(\partial_e^H\big)^{\circ n}(\underline{g}) = \alpha_v \circ \alpha_e^{\circ n}(\underline{g}). \tag{3.3}$$

To conclude the proof we define an output function $\omega : \text{Acc}^{\text{prop}} \cup K^*(\underline{f})) \to \mathcal{A}$ by

$$\omega(\underline{g}) = \begin{cases} g(e) & \text{if } \underline{g} \in \text{Acc}^{\text{prop}}(\underline{f}) \\ a & \text{otherwise,} \end{cases}$$

where $a \in \mathcal{A}$ is fixed.

Now let $\gamma = v_{i_0} H(v_{i_1}) \dots H^{\circ n}(v_{i_n})$ be the representation of γ, i.e., $v_{i_n} \neq e$. It remains to show that

$$\omega(\alpha_{v_{i_n}} \circ \dots \circ \alpha_{v_{i_0}}(\varnothing)) = \underline{f}(\gamma).$$

Since $v_{i_\gamma} \neq e$, it follows that $\partial_{v_{i_n}}^H \circ \dots \circ \partial_{v_{i_0}}^H(\underline{f}) \in \text{Acc}^{\text{prop}}$. As a consequence of Equation (3.3) one has

$$\partial_{v_{i_n}}^H \circ \dots \circ \partial_{v_{i_0}}^H(\underline{f}) = \alpha_{v_{i_n}} \circ \dots \circ \alpha_{v_{i_0}}(\underline{f}) \in \text{Acc}^{\text{prop}},$$

and therefore $\omega(\alpha_{v_{i_n}} \circ \dots \circ \alpha_{v_{i_0}}(\underline{f})) = \underline{f}(\gamma)$. This proves the assertion. $\qquad\square$

Note that the above construction of a minimal automaton is not unique. Every choice of a function Ψ gives another automaton.

Remark. Figure 3.8 is the minimal automaton one obtains from the kernel graph in Figure 3.9 using the algorithm given in the proof of Theorem 3.1.17.

Proposition 3.1.18. *Assume that $\underline{f} \in \Sigma(\Gamma, \mathcal{A})$ is (V_c, H)-automatic. If $\ker_{V_c, H}(\underline{f}) = \mathrm{Acc}^{\mathrm{prop}}(\underline{f})$, then the $\underline{f}$-automaton has $|\ker_{V_c, H}(\underline{f})|$ states.*

In other words, if every kernel element is properly accessible, then the minimal automaton is given by the kernel graph.

The kernel graph of the paperfolding sequence $\underline{pf}$ is a minimal automaton that generates $\underline{pf}$ without the property that every kernel element is properly accessible.

As we will see in the next section, if f is (V_c, H)-automatic, then f is also (W, H)-automatic, where W is a residue set for H. Therefore if W_c is another complete digit set for H, the above construction gives a minimal (W_c, H)-automaton which may or may have not more states than the minimal (V_c, H)-automaton.

Example. We consider the sequence $\bigoplus_{n \geq 0} x^n \in \Sigma(\mathbb{Z}, \{0, 1\})$, i.e., $f(n) = 1$ if and only if $n \in \mathbb{N}$, and the expanding endomorphism $H(x^j) = x^{3j}$. Then both $V_c = \{x^{-1}, x^0, x^1\}$ and $W_c = \{x^{-7}, x^0, x^1\}$ are complete digit sets for H. Then the minimal (V_c, H)-automaton for the sequence f has two states and the minimal (W_c, H)-automaton has five states. In fact, in both cases the minimal automaton is provided by the kernel graphs, see Figure 3.10 for the (V_c, H)-kernel graph and Figure 3.11 for the (W_c, H)-kernel graph.

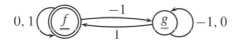

Figure 3.10. (V_c, H)-kernel graph for the sequence $\underline{f} = \bigoplus_{n \geq 0} x^n$.

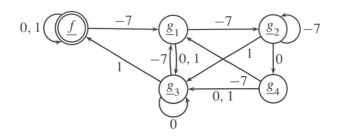

Figure 3.11. (W_c, H)-kernel graph for the sequence $\underline{f} = \bigoplus_{n \geq 0} x^n$.

We conclude this section with some remarks on generalizations of automatic sequences, i.e., (V, H)-automatic sequences for V not being a complete digit set. To this end, we have to consider properties of the remainder- and image-part-maps defined by V and H, see Definition 2.2.2. We introduce a more general concept of automatic sequences which becomes the concept of Definition 3.1.1 for complete digit sets.

To begin with, let us assume that V is a residue set for H such that $\operatorname{Per} \kappa_{H,V}$ consists of fixed points only, i.e., $\operatorname{Per} \kappa = \operatorname{Fix} \kappa$. Then Γ is the disjoint union of sets Γ_ξ with $\xi \in \operatorname{Fix} \kappa$, where

$$\Gamma_\xi = \{\gamma \mid \text{there exists } n \in \mathbb{N} \text{ such that } \kappa^{\circ n}(\gamma) = \xi\} = \bigcup_{n \in 0} \kappa^{\circ -n}(\{\xi\}).$$

For a sequence $\underline{f} \in \Sigma(\Gamma, \mathcal{A})$ we therefore have a canonical decomposition

$$\underline{f} = \bigoplus_{\xi \in \operatorname{Fix} \kappa} (\underline{f})_\xi,$$

where $(\underline{f})_\xi = \underline{f}|_{\Gamma_\xi}$.

Let $\xi \in \operatorname{Fix} \kappa$ be a fixed point of κ, then there exists a unique $v_\xi \in V$ such that $\xi = v_\xi H(\xi)$. For all $\gamma \in \Gamma \setminus \operatorname{Fix} \kappa$ there exists a unique representation

$$\gamma = v_{i_0} \ldots H^{\circ n}(v_{i_n}) H^{\circ n+1}(\xi),$$

where $n = \min\{k \mid \kappa^{\circ k+1}(\gamma) \in \operatorname{Fix} \kappa\}$, or, equivalently, $v_{i_n} \neq v_\xi$.

By analogy, we define $(\underline{f})_\xi$ to be (V, H)-automatic if there exist a finite set $\overline{\mathcal{B}}$ and maps $\alpha_v : \overline{\mathcal{B}} \to \overline{\mathcal{B}}$, $v \in V$, $\omega : \overline{\mathcal{B}} \to \mathcal{A}$ such that

1. $\omega(\varnothing) = (\underline{f})_\xi(\xi)$

2. For $\gamma \in \Gamma_\xi \setminus \{\xi\}$ with representation $\gamma = v_{i_0} \ldots H^{\circ n}(v_{i_n}) H^{\circ n+1}(\xi)$ one has

$$\omega(\alpha_{v_{i_n}} \circ \cdots \alpha_{v_{i_0}}(\varnothing)) = (\underline{f})_\xi(\gamma).$$

Taking the above considerations into account, we can extend the notion of automaticity.

Definition 3.1.19. Let $\underline{f} \in \Sigma(\Gamma, \mathcal{A})$ and let V be a residue set of H such that $\operatorname{Per} \kappa$ consists of fixed points only. Then $\underline{f}$ is called (V, H)-*automatic* if each $(\underline{f})_\xi$ is (V, H)-automatic.

It is now plain how to generalize the notion to arbitrary residue sets V. Let $\operatorname{Per} \kappa$ denote the set of periodic points, then for every $\gamma \in \Gamma \setminus \operatorname{Per} \kappa$ there exists a minimal n such that $\kappa^{\circ n}(\gamma) \in \operatorname{Per} \kappa$; this defines a map $\Xi : \Gamma \to \operatorname{Per} \kappa$ with $\Xi(\gamma) = \kappa^{\circ n}(\gamma)$. Thus Γ can be written as the disjoint union

$$\Gamma = \bigcup_{\xi \in \operatorname{Per} \kappa} \Gamma_\xi,$$

where $\Gamma_\xi = \{\gamma \mid \Xi(\gamma) = \xi\}$ and $\xi \in \operatorname{Per} \kappa$. Any sequence $\underline{f} \in \Sigma(\Gamma, \mathcal{A})$ has a decomposition

$$\underline{f} = \bigoplus_{\xi \in \operatorname{Per} \kappa} (\underline{f})_{\xi_j},$$

where $(\underline{f})_\xi = \underline{f}|_{\gamma_\xi}$. In a similar fashion as above we can now define a notion of automaticity for a sequence $\underline{f}$. This notion of automaticity is based on the fact that each $\gamma \in \Gamma \setminus \operatorname{Per} \kappa$ has a unique representation

$$\gamma = v_{i_0} H(v_{i_1}) \dots H^{\circ n}(v_{i_n}) H^{\circ n+1}(\Xi(\gamma)).$$

The restriction of $\underline{f}$ on Γ_ξ, i.e., the sequence $(\underline{f})_\xi$ is (V, H)-automatic if there exists a finite set $\overline{\mathcal{B}}$ and maps $\alpha_v : \overline{\mathcal{B}} \to \overline{\mathcal{B}}$, $\omega : \overline{\mathcal{B}} \to \mathcal{A}$ such that the following holds:

1. $\omega(\varnothing) = (\underline{f})_\xi(\xi)$.

2. For $\gamma \in \Gamma_\xi \setminus \{\xi\}$ with representation $\gamma = v_{i_0} H(v_{i_1}) \dots H^{\circ n}(v_{i_n}) H^{\circ n+1}(\xi)$ one has

$$\omega(\alpha_{v_{i_0}} \circ \dots \circ \alpha_{v_{i_0}}(\varnothing)) = (\underline{f})_{\xi_j}(\gamma).$$

Thus a sequence $\underline{f}$ is (V, H)-automatic if and only if each sequence $\underline{f}_\xi$ is automatic in the above sense.

This shows that in case of $\operatorname{Per} \kappa \neq \varnothing$ several different automata are necessary to produce a sequence $\underline{f}$. The following lemma shows that the (V, H)-kernel graph of the sequence $\underline{f}$ provides already these different automata if we allow different output functions.

Lemma 3.1.20. *Let $\underline{f} \in \Sigma(\Gamma, \mathcal{A})$ be such that the (V, H)-kernel of $\underline{f}$ is finite. For $\xi \in \operatorname{Per} \kappa$ we define $\omega_\xi : \ker_{V,H}(\underline{f}) \to \mathcal{A}$ by*

$$\omega_\xi(\underline{h}) = \underline{h}(\xi).$$

Then for each $\xi \in \operatorname{Per} \kappa$ the automaton defined by $(\ker_{V,H}, \underline{f}, \omega_\xi, \mathcal{A}, \{\partial_v^H \mid v \in V\})$ generates the sequence $(\underline{f})_\xi$ in the above sense.

The proof is easy and left to the reader.

The poor treatment of the generalized notion of automaticity is justified by the results of the following section. There we shall show that in order to study (V, H)-automatic sequences it is sufficient to study sequences and their (V, H)-kernels with respect to a complete digit set.

3.2 Elementary properties of automatic sequences

In this section we start to investigate certain elementary, but important properties of sequences generated by substitutions. One of the main results is that a sequence has a finite (V, H)-kernel if and only if the sequence has a finite $(V H(V), H^{\circ 2})$-kernel. Moreover, if a sequence has a finite (V, H)-kernel then it has a finite (W, H)-kernel for any other residue set W of H. Thus the finiteness of the kernel of a sequence is a property of the sequence and the expanding map H and does not depend on the residue set chosen. In view of these results we simply say that a sequence is H-automatic.

One should also compare the results with our considerations of the previous section. If a sequence has a finite (V, H)-kernel, then it has a finite $(W, H^{\circ n})$-kernel, where W is a residue set of $H^{\circ n}$ and $n \in \mathbb{N}$. By Theorem 2.2.7, we know that for n sufficiently large we always can find a complete digit set W_c for $H^{\circ n}$. Thus the concept of automaticity introduced for complete digit sets is sufficient for a study of automatic sequences.

However, although the automaticity of a sequence does not depend on the residue set chosen it is often advantageous to choose a certain residue set, e.g., if one wants to compute the substitution associated with an automatic sequence. Moreover, as we have seen, the minimal automaton that generates a sequence depends on the residue set.

Theorem 3.2.1. *Let $\underline{f} \in \Sigma(\Gamma, \mathcal{A})$, then $\underline{f}$ has a finite (V, H)-kernel if and only if $\underline{f}$ has a finite $(V H(V), H^{\circ 2})$-kernel.*

Proof. Let $\ker_{V,H}(\underline{f})$ be finite. By 1. of Corollary 2.2.13, we have for $\tau = v H(w) \in V H(V)$

$$\partial_\tau^{H^{\circ 2}} = \partial_{v H(w)}^{H^{\circ 2}} = \partial_w^H \circ \partial_v^H$$

for all $v, w \in V$. In other words, $\partial_\tau^{H^{\circ 2}}(\underline{f}) \in \ker_{V,H}$ for all $\tau \in V H(V)$. This means $\ker_{V H(V), H^{\circ 2}}(\underline{f}) \subset \ker_{V,H}(\underline{f})$. Thus the finiteness of the (V, H)-kernel of $\underline{f}$ implies the finiteness of the $(V H(V); H^{\circ 2})$-kernel of $\underline{f}$.

Now, we assume that the $(V H(V), H^{\circ 2})$-kernel of $\underline{f}$ is finite. Then the set

$$\overline{K} = \ker_{V H(V), H^{\circ 2}} \cup \bigcup_{v \in V} \partial_v^H \left(\ker_{V H(V), H^{\circ 2}}(\underline{f}) \right)$$

contains $\underline{f}$ and is finite. In order to show that the (V, H)-kernel of $\underline{f}$ is finite it suffices to show that $\overline{K}$ is invariant under v-decimations, $v \in V$, i.e., $\partial_v^H(\overline{K}) \subset \overline{K}$; this is a simple consequence of the definition of $\overline{K}$. $\qquad\square$

Similar arguments show that the same assertion is true for higher iterates $H^{\circ n}$ of H and the corresponding residue sets $\kappa^{\circ -n}(V)$.

For several of the following results we need the fact that balls with radius $r > 0$ and center $\gamma \in \Gamma$ contain only finitely many elements. This is guaranteed if $\| \ \|$ on Γ is discrete, see Definition 2.1.7.

The next theorem shows that the finiteness of the (V, H)-kernel of $\underline{f}$ is independent of the residue system.

Theorem 3.2.2. *Let* $\| \ \|$ *be a discrete norm on* Γ *and let* $H : \Gamma \to \Gamma$ *be expanding with* $H(\Gamma)$ *of finite index in* Γ. *Let* $\underline{f} \in \Sigma(\Gamma, \mathcal{A})$ *and let* $\ker_{V,H}(\underline{f})$ *be finite. If* W *is a residue system for* H, *then* $\ker_{W,H}(\underline{f})$ *is finite, too.*

Proof. Let $r = \max\{\|v\| \mid v \in V\}$ and $R = \max\{\|\kappa_{H,V}(w)\| \mid w \in W\}$ and let C be the expansion ratio of H. We shall show that $\ker_{W,H}(\underline{f}) \subset \overline{K}$, where $\overline{K}$ is defined by

$$\overline{K} = \left\{ (T_\gamma)^*(\underline{h}) \mid \underline{h} \in \ker_{V,H}(\underline{f}) \text{ and } \|\gamma\| \leq \tfrac{2r+CR}{C-1} \right\}.$$

Since $\ker_{V,H}(\underline{f})$ and $B_{\frac{2r+CR}{C-1}}(e)$ are finite sets, $\overline{K}$ is a finite set and therefore the invariance of $\overline{K}$ under ∂_w^H-decimations, $w \in W$, would imply that $\ker_{W,H}(\underline{f})$ is finite.

By the definition of $\overline{K}$, we have $\underline{f} \in \overline{K}$. Therefore, it suffices to prove that $\partial_w^H(\overline{K}) \subset \overline{K}$ for all $w \in W$.

To this end, we write

$$\partial_w^H = (T_{\kappa_{H,V}(w)})^* \circ \partial_{\zeta_{H,V}(w)}^H,$$

see Lemma 2.2.14. Furthermore, let $\gamma \in \Gamma$ such that $\|\gamma\| \leq \frac{2r+CR}{C-1}$ and consider for $w \in W$

$$\partial_w^H \circ (T_\gamma)^* = (T_{\kappa_{H,V}(w)})^* \circ \partial_{\zeta_{H,V}(w)}^H \circ (T_\gamma)^*.$$

By Lemma 2.2.15 this gives

$$(T_{\kappa_{H,V}(w)})^* \circ (T_{\kappa_{H,V}(\gamma\zeta_{H,V}(w))})^* \circ \partial_{\zeta_{H,V}(\gamma\zeta_{H,V}(w))}^H$$
$$= (T_{\kappa_{H,V}(\gamma\zeta_{H,V}(w))\kappa_{H,V}(w)})^* \circ \partial_{\zeta_{H,V}(\gamma\zeta_{H,V}(w))}^H.$$

We now estimate

$$\|\kappa_{H,V}(\gamma\zeta_{H,V}(w))\kappa_{H,V}(w)\| \leq \|\kappa_{H,V}(\gamma\zeta_{H,V}(w))\| + \|\kappa_{H,V}(w)\|.$$

From the proof of Lemma 2.2.3 we obtain

$$\begin{aligned}
\|\kappa_{H,V}(\gamma\zeta_{H,V}(w))\| + \|\kappa_{H,V}(w)\| &\leq \frac{\|\gamma\zeta_{H,V}(w)\| + r}{C} + R \\
&\leq \frac{\|\gamma\| + 2r}{C} + R \\
&\leq \frac{2r + CR}{C - 1}
\end{aligned}$$

by our choice of γ. Therefore $\partial_w^H(\overline{K}) \subset \overline{K}$.   □

From now on it is justified to speak of the H-kernel of f which we denote by $\ker_H(\underline{f})$. Furthermore, we may restate Theorems 2.2.19 and 3.1.6

Theorem 2.2.19*. *The sequence $\underline{f} \in \Sigma(\Gamma, \mathcal{A})$ is generated by a substitution (w.r.t. H) if and only if the H-kernel of $\underline{f}$ is finite.*

Theorem 3.1.6*. *The sequence $\underline{f} \in \Sigma(\Gamma, \mathcal{A})$ is H-automatic if and only if the H-kernel of $\underline{f}$ is finite.*

The next lemma shows that the change of the residue set does not create entirely new sequences. In fact, the sequences contained in $\ker_{W,H}(\underline{f})$ are translations of the elements in $\ker_{V,H}(\underline{f})$.

Lemma 3.2.3. *Let $\underline{f}$ be a sequence with a finite H-kernel. If V and W are residue sets for H and $\ker_V(\underline{f})$, $\ker_W(\underline{f})$ are the kernels w.r.t. V and W, respectively, then the following holds.*

1. *For any $\underline{h} \in \ker_V(\underline{f})$ there exists a $\rho \in \Gamma$ such that $(T_\rho)^*(\underline{h}) \in \ker_W(\underline{f})$.*

2. *For any $\underline{g} \in \ker_W(\underline{f})$ there exists a $\underline{h} \in \ker_V(\underline{f})$ and a $\rho \in \Gamma$ such that $\underline{g} = (T_\rho)^*(\underline{h})$.*

Proof. Let $\underline{h} = \partial_{v_1}^H \circ \cdots \circ \partial_{v_k}^H(\underline{f}) \in \ker_V(\underline{f})$, where $v_j \in V$, $j = 1, \ldots, k$. By Lemma 2.2.14, we can replace each $\partial_{v_j}^H$ by $(T_{\kappa_W(v_j)})^* \circ \partial_{\zeta_W(v_j)}^H$, where $\zeta_W(v_j) \in W$. This gives

$$\underline{h} = ((T_{\kappa_W(v_1)})^* \circ \partial_{\zeta_W(v_1)}^H) \circ \cdots \circ (T_{\kappa_W(v_k)})^* \circ \partial_{\zeta_W(v_k)}^H)(\underline{f}).$$

A repeated application of Lemma 2.2.15 yields

$$\underline{h} = (T_\rho)^* \circ \partial_{w_1'}^H \circ \cdots \circ \partial_{w_k'}^H(\underline{f})$$

for a certain $\rho \in \Gamma$ and $w_j' \in W$, $j = 1, \ldots, k$. This proves the first assertion.
    The second assertion follows from the above proof by interchanging the roles of V and W.   □

The next lemma relates H-automatic sequences with certain G-automatic sequences.

Lemma 3.2.4. *Let $G, H : \Gamma \to \Gamma$ be expanding endomorphisms. Suppose there exist $k, n \in \mathbb{N}$ such that $H^{\circ n} = G^{\circ k}$. Then $\underline{f} \in \Sigma(\Gamma, \mathcal{A})$ has a finite H-kernel if and only if $\underline{f}$ has a finite G-kernel.*

Proof. The proof is a direct consequence of Theorem 3.2.1 and Theorem 3.2.2.   □

As a next step, we are interested in the following question. Suppose that f is an H-automatic sequence. What are possible alterations of the sequence which preserve the automaticity? In the remaining part of this section we will give some preliminary answers. In the next chapter we shall investigate this question in greater detail.

We start with a translation invariance property.

Theorem 3.2.5. *Let $\| \ \|$ be discrete on Γ. If $f \in \Sigma(\Gamma, \mathcal{A})$ has a finite H-kernel, then $(T_\rho)^*(f)$ has a finite H-kernel.*

Proof. Let V be a residue set of H and let $r = \max\{\|v\| \mid v \in V\}$. Let C be the expansion ratio of H and let the finite set $\overline{K}$ be defined by

$$\overline{K} = \left\{ (T_\gamma)^*(\underline{h}) \mid \|\gamma\| \leq \max\left\{\|\rho\|, \tfrac{2r}{C-1}\right\} \text{ and } \underline{h} \in \ker_{V,H}(f) \right\}.$$

As usual, we shall prove that $\ker_{V,H}((T_\rho)^*(f)) \subset \overline{K}$. By construction, we have that $(T_\rho)^*(f) \in \overline{K}$. It is therefore sufficient to show that $\partial_v^H(\overline{K}) \subset \overline{K}$ for all $v \in V$.

Let $\gamma \in \Gamma$ such that $\|\gamma\| \leq \max\{\|\rho\|, \tfrac{2r}{C-1}\}$ and let $v \in V$, then by Lemma 2.2.15

$$\partial_v^H \circ (T_\gamma)^* = (T_{\kappa(\gamma v)})^* \circ \partial_{\zeta(\gamma v)}^H.$$

Again, we estimate

$$\|\kappa(\gamma v)\| \leq \frac{\|\gamma v\| + r}{C} \leq \frac{\|\gamma\| + 2r}{C} \leq \max\left\{\|\rho\|, \frac{2r}{C-1}\right\}.$$

This proves the assertion.   □

Remarks.

1. Since $(T_\rho)^* = (T_{\rho^{-1}})_*$ the same assertion holds for $(T_\rho)_*(f)$.

2. The above lemma does not hold for right translations, as the example below shows. The lemma is true for right translations $R_\rho(\gamma) = \gamma\rho$ under an additional assumption on the group Γ, namely if the commutators

$$[H(\gamma), \rho] = H(\gamma^{-1})\rho^{-1}H(\gamma)\rho$$

    are uniformly bounded, i.e., there exists a constant K such that $\|[H(\gamma), \rho]\| \leq K$ for all ρ and all γ. This seems to be a rather strong condition on the group Γ which is certainly satisfied by commutative groups.

The Heisenberg group provides an example for which H-automaticity is not preserved under right translations.

Example. Let Γ_3 be the Heisenberg group and $H_{2,2} : L \to L$ be as in Example 3, p. 28. Then $V = \{b^{\epsilon_1} a^{\epsilon_2} c^{\epsilon_3} \mid \epsilon_1, \epsilon_2 \in \{0, 1\}, \epsilon_3 \in \{0, 1, 2, 3\}\}$ is a residue set for $H_{2,2}$. Now consider the sequence $\underline{f} \in \Sigma(\Gamma_3, \{0, 1\})$ defined by $\underline{f}(\gamma) = 1$ if and only if $\gamma = b^n$ with $n \in \mathbb{N}$. Then $\underline{f}$ has a finite $H_{2,2}$-kernel. Indeed, we have

$$\partial_v^{H_{2,2}}(\underline{f}) = \underline{0} \quad \text{for all } v \neq e, b, b^2,$$

$$\partial_v^{H_{2,2}}(\underline{f}) = \underline{f} \quad v \in \{e, b, b^2\}.$$

Due to the above theorem, we have that the sequence $(T_{a^{-1}})^*(\underline{f})$ is H-automatic. If $R_{a^{-1}} : \Gamma_3 \to \Gamma_3$ denotes the right translation, then the sequence $(R_{a^{-1}})^*(\underline{f})$ is not H-automatic.

In order to prove the non-automaticity of $(R_{a^{-1}})^*(\underline{f})$ we introduce the set $M_1(\underline{h})$ for an arbitrary sequence $\underline{h} \in \Sigma(\Gamma_3, \{0, 1\})$. It is defined by

$$M_1(\underline{h}) = \{\gamma \mid \underline{h}(\gamma) = 1\}.$$

By definition, we have $(R_{a^{-1}})^*(\underline{f})(\gamma) = \underline{f}(\gamma a^{-1})$, so we obtain $(R_{a^{-1}})^*(\underline{f})(\gamma) = 1$ if and only if $\gamma = b^n a = ab^n c^n$ and therefore

$$M_1((R_{a^{-1}})^*(\underline{f})(\gamma)) = \{ab^n c^n \mid n \in \mathbb{N}\}.$$

Now consider $\partial_a((R_{a^{-1}})^*(\underline{f}))(\gamma) = (R_{a^{-1}})^*(\underline{f})(aH(\gamma))$, and we thus obtain that $\partial_a((R_{a^{-1}})^*(\underline{f}))(\gamma) = 1$ if and only if $aH(\gamma) = ab^n c^n$, i.e., $H(\gamma) = b^n c^n$. This is only possible if n is a multiple of 4 which yields

$$M_1(\partial_a((R_{a^{-1}})^*(\underline{f}))) = \{b^{2n} c^n \mid n \in \mathbb{N}\}.$$

Finally we compute the set

$$M_1(\partial_e(\partial_a((R_{a^{-1}})^*(\underline{f})))).$$

By our above considerations we see that

$$\partial_e(\partial_a((R_{a^{-1}})^*(\underline{f})))(\gamma) = 1$$

if and only if $H(\gamma) \in \{b^{2n} c^n \mid n \in \mathbb{N}\}$, i.e., $\gamma = b^{4n} c^n$ and $n \in \mathbb{N}$. By induction, we obtain

$$M_1(\partial_e^{\circ k}(\partial_a((R_{a^{-1}})^*(\underline{f})))) = \{b^{2^{k+1}n} c^n \mid n \in \mathbb{N}\}.$$

This proves that the H-kernel of $(R_{a^{-1}})^*(\underline{f})$ is infinite.

The next theorem is about the finiteness of the kernel of a G-reduction of $\underline{f}$.

Theorem 3.2.6. *Let $\| \ \|$ be discrete on Γ. Let $G : \Gamma \to \Gamma$ be a group endomorphism which commutes with H, i.e., $H \circ G = G \circ H$. If the H-kernel of $\underline{f}$ is finite, then the H-kernel of $G^*(\underline{f})$ is finite.*

Proof. Let V be a residue system for H (with expansion ratio C) and let $r = \max\{\|v\| \mid v \in V\}$, $\tilde{r} = \max\{\|G(v)\| \mid v \in V\}$. Define the finite set $\overline{K}$ by

$$\overline{K} = \left\{ G^* \circ (T_\gamma)^*(\underline{h}) \mid \underline{h} \in \ker_{V,H}(\underline{f}) \text{ and } \|\gamma\| \leq \tfrac{\tilde{r}+r}{C-1} \right\}.$$

We thus have $G^*(\underline{f}) \in \overline{K}$, and it suffices to show the invariance of $\overline{K}$ under the v-decimations. Let $\|\gamma\| \leq \tfrac{\tilde{r}+r}{C-1}$, then we have

$$
\begin{aligned}
\partial_v^H \circ G^* \circ (T_\gamma)^* &= (T_v \circ H)^* \circ (T_\gamma \circ G)^* \\
&= (T_\gamma \circ G \circ T_v \circ H)^* \\
&= (T_\gamma \circ T_{G(v)} \circ G \circ H)^* \\
&= (T_\gamma \circ T_{G(v)} \circ H \circ G)^* \qquad \text{since } H \circ G = G \circ H \\
&= G^* \circ \partial_{\gamma G(v)}^H \\
&= G^* \circ (T_{\kappa_{H,V}(\gamma G(v))})^* \circ \partial_{\zeta_{H,V}(\gamma G(v))}^H
\end{aligned}
$$

which gives

$$\|\kappa_{H,V}(\gamma G(v))\| \leq \frac{\|\gamma G(v)\| + r}{C} \leq \frac{\|\gamma\| + r + \tilde{r}}{C} \leq \frac{\tilde{r}+r}{C-1}$$

due to our choice of γ. $\qquad \square$

If $G : \Gamma \to \Gamma$ commutes with H and if G is not surjective, then Theorem 3.2.6 can be formulated as follows: A carefully chosen subsequence of $\underline{f}$, i.e., the sequence $G^*(\underline{f})$, of an H-automatic sequence $\underline{f}$ is H-automatic, too.

If G is a bijection and commutes with H, then we can say that a careful rearrangement, i.e., $G^*(\underline{f})$, of an H-automatic sequence is also H-automatic.

One special rearrangement is the endomorphism $G(\gamma) = \gamma^{-1}$.

Proposition 3.2.7. *Let* $G(\gamma) = \gamma^{-1}$. *Then* $G \circ H = H \circ G$, *and therefore* $\underline{f}$ *has a finite* H-*kernel if and only if this holds for* $G^*(\underline{f})$.

A slight modification of the proof of Theorem 3.2.6 yields the following generalization of Theorem 3.2.6.

Corollary 3.2.8. *Let* $\{G_1, \ldots, G_M\}$ *be a finite set of group endomorphisms of* Γ *and let* $\eta : \{1, \ldots, M\} \to \{1, \ldots, M\}$ *be a map such that*

$$G_j \circ H = H \circ G_{\eta(j)}$$

holds for all $j \in \{1, \ldots, M\}$. *If the* H-*kernel of* $\underline{f}$ *is finite, then so is the* H-*kernel of* $(G_j)^*(\underline{f})$ *for all* $j \in \{1, \ldots, M\}$.

Example. Let Γ be the additive group $\mathbb{Z}^2$. Then $H : \mathbb{Z}^2 \to \mathbb{Z}^2$ defined by

$$H \begin{pmatrix} x \\ y \end{pmatrix} = \begin{pmatrix} x - y \\ x + y \end{pmatrix}$$

is an expanding endomorphism (w.r.t. the euclidian metric). Let $G_0 : \mathbb{Z}^2 \to \mathbb{Z}^2$ be defined by

$$G_0 \begin{pmatrix} x \\ y \end{pmatrix} = \begin{pmatrix} a & b \\ c & d \end{pmatrix} \begin{pmatrix} x \\ y \end{pmatrix},$$

where $a + b + c + d = 0 \bmod 2$. Then the set $\{H^{\circ -n} G_0 H^{\circ n} \mid n \in \mathbb{N}\}$ is finite and satisfies the requirements of Corollary 3.2.8.

As another application of Theorem 3.2.6 we show that a kind of blocking of an automatic sequence again yields an automatic sequence.

The blocking is best explained by an example. If $(f_n)_{n \geq 0}$ is a sequence with values in $\mathcal{A}$, then the sequence $(g_n)_{n \geq 0}$ defined by $g_n = (f_{2n}, f_{2n+1})$ is a new sequence with values in $\mathcal{A}^2$.

Definition 3.2.9. Let $G : \Gamma \to \Gamma$ be a group endomorphism such that $G(\Gamma) \subset \Gamma$ is a subgroup of finite index and let $W = \{w_1 = e, w_2, \ldots, w_D\}$ be a residue set for G. The map $B_{W,G} : \Sigma(\Gamma, \mathcal{A}) \to \Sigma(\Gamma, \mathcal{A}^D)$ defined by

$$B_{W,G}(\underline{f})(\gamma) = \begin{pmatrix} \underline{f}(G(\gamma)) \\ \underline{f}(w_2 G(\gamma)) \\ \vdots \\ \underline{f}(w_D G(\gamma)) \end{pmatrix}$$

is called (W, G)-block map.

Lemma 3.2.10. *Let* $G : \Gamma \to \Gamma$ *be as in Definition* 3.2.9 *and assume that* G *commutes with* H, *i.e.,* $G \circ H = H \circ G$. *If* $\underline{f} \in \Sigma(\Gamma, \mathcal{A})$ *is* H-*automatic, then* $B_{W,G}(\underline{f}) \in \Sigma(\Gamma, \mathcal{A}^D)$ *is* H-*automatic.*

Proof. The proof follows immediately from the fact that the j-th component of $B_{W,G}(\underline{f})$, $j = 1, \ldots, D$, is $\partial_{w_j}^G(\underline{f}) = (T_{w_j} \circ G)^*(\underline{f})$. By Theorem 3.2.5 and Theorem 3.2.6, each component is H-automatic and therefore $B_{W,G}(\underline{f})$ is H-automatic. $\square$

It remains to study the automaticity properties of $G_*(\underline{f})$, where $G : \Gamma \to \Gamma$ is injective. As for the case $G^*(\underline{f})$ we need that G satisfies certain requirements.

Lemma 3.2.11. *Let* $G : \Gamma \to \Gamma$ *be an injective group endomorphism such that the subgroup* $G(\Gamma) \subset \Gamma$ *has finite index. Furthermore, assume that* G *commutes with* H, *i.e., $H \circ G = H \circ G$. If V is a residue set for H and W is a residue set for G, then for all $\underline{f} \in \Sigma(\Gamma, \overline{\mathcal{A}})$ and all $v \in V$*

$$\partial_v^H \circ G_*(\underline{f}) = \bigoplus_{w \in W(v)} (T_w \circ G)_* \circ (T_{\kappa(G^{-1}(vH(w)))})^* \circ \partial_{\zeta(G^{-1}(vH(w)))}^H (\underline{f}),$$

where $W(v) = \{w \mid w \in W \text{ and } vH(w) \in G(\Gamma)\}$ and κ and ζ are the image-part- and remainder-maps w.r.t. V, respectively.

Proof. By its definition, we have

$$\partial_v^H (G_*(\underline{f}))(\gamma) = G_*(\underline{f})(vH(\gamma)) = \begin{cases} \underline{f}(\rho) & \text{if } vH(\gamma) = G(\rho) \\ \varnothing & \text{otherwise.} \end{cases}$$

Now let $\gamma \in \Gamma$ be such that $vH(\gamma) \in G(\Gamma)$. Since W is a residue set of G, γ has a unique representation $\gamma = wG(\gamma')$ with $w \in W$. This yields $vH(wG(\gamma')) \in G(\Gamma)$, and therefore $vH(w)G(H(\gamma')) \in G(\Gamma)$ which implies $vH(w) \in G(\Gamma)$. Due to the definition of $W(v)$ the set

$$\bigcup_{w \in W(v)} wG(\Gamma)$$

contains all γ such that $vH(\gamma) \in G(\Gamma)$. In other words, we have

$$\partial_v^H (G_*(\underline{f}))(wG(\gamma)) = \underline{f}(vH(w)H(G(\gamma)))$$

for all $\gamma \in \Gamma$ and $w \in W(v)$. For all $\rho \in \Gamma$ which are not of the form $wG(\gamma)$ we have that $\partial_v^H (G_*(\underline{f}))(\rho) = \varnothing$. Written as a formal series we obtain

$$\partial_v^H (G_*(\underline{f})) = \bigoplus_{w \in W(v), \gamma \in \Gamma} \underline{f}(vH(w)H(G(\gamma))) \, wG(\gamma).$$

By the fact that $G^{-1}(vH(w)H(G(\gamma)))$ is well defined and equal to $G^{-1}(vH(w))H(\gamma)$ we can rewrite the summands as

$$\partial_v^H (G_*(\underline{f})) = \bigoplus_{w \in W(v)} (T_w \circ G)_* \partial_{G^{-1}(vH(w))}^H (\underline{f})$$

and an application of Lemma 2.2.14 proves the assertion. $\qquad \square$

Theorem 3.2.12. *Let $G : \Gamma \to \Gamma$ be as in Lemma 3.2.11. If $\underline{f} \in \Sigma(\Gamma, \mathcal{A})$ is H-automatic, then $G_*(\underline{f})$ is H-automatic, too.*

Proof. Let V be a residue set for H and W be a residue set for G. By Lemma 3.2.11, we have that

$$\partial_v^H (G_*(\underline{f})) = \bigoplus_{w \in W(v)} (T_w \circ G)_* \circ (T_{\kappa(G^{-1}(vH(w)))})^* \circ \partial_{\zeta(G^{-1}(vH(w)))}^H (\underline{f}).$$

Now $(T_w \circ G)_* = (T_{w^{-1}})^* \circ G_*$, and we obtain

$$\partial_v^H (G_*(\underline{f})) = \bigoplus_{w \in W(v)} (T_{w^{-1}})^* \circ G_* \circ (T_{\kappa(G^{-1}(vH(w)))})^* \circ \partial_{\zeta(G^{-1}(vH(w)))}^H (\underline{f}).$$

Note that the summands are of the form $(T_\gamma) \circ G_* \circ T_\rho^*(\underline{h})$, where $\underline{h} \in \ker_{V,H}(\underline{f})$. As a next step, we define several constants, namely

$$r = \max\{\|v\| \mid v \in V\},$$
$$R_1 = \max\{\|G^{-1}(vH(W))\| \mid v \in V \text{ and } w \in W(v)\},$$
$$R_2 = \max\{\|w\| \mid w \in W\}.$$

The set of sequences

$$K' = \left\{ T_\gamma^* \circ G_* \circ T_\rho^*(\underline{h}) \mid \underline{h} \in \ker_H(\underline{f}), \|\gamma\| \le \tfrac{CR_2+2r}{C-1}, \|\rho\| \le \tfrac{R_1+3r}{C-1} \right\},$$

where C denotes the expansion ratio of H, is a finite set. Unfortunately, it is not decimation invariant. If we apply Lemma 3.2.11 to an element of K', we obtain the following formula

$$\partial_v^H (T_\gamma^* \circ G_* \circ T_\rho^*(\underline{h})$$
$$= \bigoplus_{w \in W(\zeta(\gamma v))} T_{w^{-1}\kappa(\gamma v)}^* \circ G_* \circ T_{\kappa(\rho\zeta(G^{-1}(vH(w))\kappa(G^{-1}(vH(w)))}^* \circ \partial_{\zeta(\rho\zeta(G^{-1}(vH(w)))}^H (\underline{h})$$

and we estimate

$$\|w^{-1}\kappa(\gamma v)\| \le R_2 + \frac{\|\gamma\| + 2r}{C} \le \frac{CR_2 + 2r}{C-1},$$

due to the choice of γ, and

$$\|\kappa(\rho\zeta(G^{-1}(vH(w))\kappa(G^{-1}(vH(w)))\| \le \frac{\|\rho\| + 2r + R_1 + r}{C} \le \frac{R_1 + 3r}{C-1}$$

due to the choice of ρ. Therefore the decimation of an element in K' is a finite sum of elements in K'. Note that the summands have pairwise disjoint support. Note further that the set of sums of elements in K' where the summands have pairwise disjoint support is also a finite set.

Since $G_*(\underline{f})$ is an element of K', its decimations are well defined and can be written as a finite sum of elements of K' such that the summands have pairwise disjoint support. Therefore the H-kernel of $G_*(\underline{f})$ is finite. $\square$

Corollary 3.2.13. *Let $G : \Gamma \to \Gamma$ be as in Lemma 3.2.11 and let $s : W \times \mathcal{A} \to \mathcal{B}$ be a map, where $\mathcal{A}$ and $\mathcal{B}$ are finite sets. Then S induces a kind of (W, G)-substitution S from $\Sigma(\Gamma, \mathcal{A})$ to $\Sigma(\Gamma, \mathcal{B})$ by setting*

$$S(\underline{f})(wG(\gamma)) = s(w, \underline{f}(\gamma)).$$

If $\underline{f} \in \Sigma(\Gamma, \mathcal{A})$ is H-automatic, then $S(\underline{f})$ is also H-automatic.

Proof. We have

$$S(\underline{f}) = \bigoplus_{w \in W} \hat{s_w}((T_w \circ G)_*(\underline{f})),$$

where $s_w : \mathcal{A} \to \mathcal{B}$ denotes the map $s(w, \ )$. By the above theorem and Theorem 3.2.5 each of the summands is H-automatic. Therefore the sum is H-automatic. $\square$

The above results indicate that it is possible to perform alterations of an H-automatic sequence without destroying the H-automaticity of the sequence. So far, we have seen that as long as the alterations are compatible with the group structure and with the expanding endomorphism H, then the H-automaticity is preserved.

On the other hand, there are simple alterations of a sequence $\underline{f}$ which certainly do not affect the automaticity of the sequence. One of the simplest of these changes is the exchange of two values. Let $\underline{f}$ be an H-automatic sequence and let γ_1, γ_2 be different elements of Γ, then $\underline{g}$ defined by

$$\underline{g}(\gamma) = \begin{cases} \underline{f}(\gamma_1) & \text{if } \gamma = \gamma_2 \\ \underline{f}(\gamma_2) & \text{if } \gamma = \gamma_1 \\ \underline{f}(\gamma) & \text{otherwise} \end{cases}$$

is an H-automatic sequence, too. Unfortunately, this kind of exchange of two values is not covered by mappings $G : H \to H$ which satisfy one of the above mentioned conditions. It is clear that any finite number of changes of an H-automatic sequence $\underline{f}$ produces a sequence $\underline{g}$ which is H-automatic.

There is yet another, more complicated, way to rearrange an automatic sequence. Let us assume that $(\overline{\mathcal{B}}, \varnothing, \mathcal{A}, \omega, \{\alpha_v \mid v \in V_c\})$ is a (V_c, H)-automaton generating the sequence $\underline{f} \in \Sigma(\Gamma, \mathcal{A})$. If $\sigma : V_c \to V_c$ is bijective and $\sigma(e) = e$, then the automaton $(\overline{\mathcal{B}}, \varnothing, \mathcal{A}, \omega, \{\alpha_{\sigma(v)} \mid v \in V_c\})$ generates a sequence which can be considered as a rearrangement of $\underline{f}$. By the construction, it is clear that the new sequence is also automatic.

In the next chapter we shall introduce the necessary tools to deal with more general rearrangements of sequences.

3.3   Notes and comments

The definition of a (V_c, H)-automaton is a special case of the general notion of a finite automaton. More information on finite automata and their relations to other structures in mathematics and computer science can be found in, e.g., [45], [48], [77], [78], [85], [98], [105], [124], [147], [156], [159].

Definition 3.1.1 is a generalization of the concept of automaticity for 'true' sequences, i.e., maps from $\mathbb{N}$ to a set. In [7], [9], [15], [17], [153], the reader will find further generalizations of automaticity.

Theorem 3.1.6 is a generalization of a result in [55] for true sequences.

Chapter 4

Automaticity II

As we have seen in the previous section there exist operations on the set of H-automatic sequences which preserve the H-automaticity. We have also seen that the allowed operations are very restrictive since they depend on the group structure as well as on the expanding endomorphism H. In this chapter we develop a more general theory of admissible alterations of H-automatic sequences.

We begin with H-automatic subsets of Γ. We study the behavior of H-automatic subsets under the elementary set operations.

Using the notion of automatic subsets, we define $(H_1 \times H_2)$-automatic maps $G : \Gamma_1 \to \Gamma_2$, where $H_i : \Gamma_i \to \Gamma_i$, $i = 1, 2$, are expanding endomorphisms such that the index of $H(\Gamma_i)$ is finite. We study the properties of automatic maps. In particular, we shall show that an $(H_1 \times H_2)$-automatic map $G : \Gamma_1 \to \Gamma_2$ maps an H_1-automatic subset $M \subset \Gamma_1$ on the H_2-automatic subset $G(M)$.

In the third section, we discuss the structure of finite $(V_1 \times V_2, H_1 \times H_2)$-automata that generate $(H_1 \times H_2)$-automatic maps $G : \Gamma_1 \to \Gamma_2$. We present necessary and sufficient conditions on the transition graph of a $(V_1 \times V_2, H_1 \times H_2)$-automaton to generate an $(H_1 \times H_2)$-automatic map. We conclude this section by showing that there exists an $(H \times H_p)$-automatic bijective map $\mathfrak{V}_p : \Gamma \to \mathbb{N}$ such that for every H-automatic subset $M \subset \Gamma$ the set $\mathfrak{V}_p(M) \subset \mathbb{N}$ is H_p-automatic, where $H_p(x^j) = x^{pj}$ and $p = |V|$.

In the fourth section we present a detailed discussion of the properties of $(H_p \times H_q)$-automatic functions $G : \mathbb{N} \to \mathbb{N}$. We present upper and lower bounds for the growth of $(H_p \times H_q)$-automatic functions. Moreover, we prove that for $p < q$ there exists no surjective $(H_p \times H_q)$-automatic function $G : \mathbb{N} \to \mathbb{N}$. By studying the distribution or density of an H_q-automatic subset we are able to give sufficient conditions for the existence of an $(H_p \times H_q)$-automatic function $G : \mathbb{N} \to \mathbb{N}$ such that $p < q$ and $G(\mathbb{N}) = M$.

Finally, in the fifth section, we briefly introduce cellular automata and show that under mild restrictions a cellular automaton is a tool to generate an $(H_1 \times H_2)$-automatic map $G : \Gamma_1 \to \Gamma_2$.

4.1 Automatic subsets

As always, Γ is a finitely generated group and $H : \Gamma \to \Gamma$ is an expanding endo-morphism (w.r.t. some fixed discrete norm) such that $H(\Gamma)$ is of finite index in Γ. Moreover, $\mathcal{A}$ denotes a finite set and $\Sigma(\Gamma, \mathcal{A})$ denotes the set of sequences.

In this section, we discuss the notion of automatic subsets of Γ. It will turn out that a study of automatic subsets almost suffices to understand arbitrary H-automatic sequences.

Definition 4.1.1. A subset $M \subset \Gamma$ is called H-*automatic* if the characteristic sequence $\chi_M \in \Sigma(\Gamma, \{0, 1\})$ defined by

$$\chi_M(\gamma) = \begin{cases} 1 & \text{if } \gamma \in M \\ 0 & \text{otherwise} \end{cases}$$

is H-automatic.

Examples.

1. The empty set $\emptyset$ and the set Γ are H-automatic subsets. Indeed, we have $\chi_\emptyset(\gamma) = 0$ and $\chi_\Gamma(\gamma) = 1$ for all $\gamma \in \Gamma$. Therefore $\partial_v^H(\chi_\emptyset) = \chi_\emptyset$ and $\partial_v^H(\chi_\Gamma) = \chi_\Gamma$ for all $v \in V$.

   From now on $\chi_\emptyset$ is simply denoted by $\underline{0}$, the sequence with values all equal to 0.

2. Every finite subset M of Γ is an H-automatic subset.

3. If M is equal to $\Gamma \setminus \{\gamma_1, \ldots, \gamma_n\}$, where $n \in \mathbb{N}$, then M is an H-automatic subset.

In order to model the different operations with subsets, we introduce an additional structure on the set $\{0, 1\}$. We denote this structure by $\mathbb{B} = (\{0, 1\}, \vee, \wedge, \bar{\ })$, i.e., we consider the set $\{0, 1\}$ as a two element Boolean algebra, where

$$0 \vee 0 = 0 \quad \text{and} \quad 1 \vee 0 = 0 \vee 1 = 1 \vee 1 = 1,$$

and

$$1 \wedge 1 = 1 \quad \text{and} \quad 1 \wedge 0 = 0 \wedge 1 = 0 \wedge 0 = 0,$$

and $0^- = 1, 1^- = 0$.

The operations given for the Boolean algebra can be extended to the sequence space $\Sigma(\Gamma, \mathbb{B})$. If $\underline{f} \in \Sigma(\Gamma, \mathbb{B})$, then $\underline{f}^-$ is defined by

$$\underline{f}^-(\gamma) = \underline{f}(\gamma)^-.$$

For two given sequences $\underline{f}, \underline{g} \in \Sigma(\Gamma, \mathbb{B})$ we define their sum by

$$(\underline{f} \vee \underline{g})(\gamma) = \underline{f}(\gamma) \vee \underline{g}(\gamma).$$

As a formal series we write a sequence $\underline{f} \in \Sigma(\Gamma, \mathbb{B})$ as

$$\underline{f} = \bigvee_{\gamma \in \Gamma} f_\gamma \, \gamma$$

to emphasize the addition on $\mathbb{B}$. There are to different kinds of products. The first one is a pointwise product, it is induced by $\wedge$ and defined by

$$(\underline{f} \wedge \underline{g})(\gamma) = \underline{f}(\gamma) \wedge \underline{g}(\gamma),$$

or, in the notion of formal series,

$$\underline{f} \wedge \underline{g} = \bigvee_{\gamma \in \Gamma} (f_\gamma \wedge g_\gamma) \, \gamma.$$

The second product is the so-called Cauchy product. It is defined by

$$(\underline{f} \cdot \underline{g})(\gamma) = \bigvee_{\rho \tau = \gamma} f_\rho \cdot g_\tau,$$

where the summation is over all pairs (ρ, τ) with $\rho \tau = \gamma$, or, in the notion of formal series,

$$(\underline{f} \cdot \underline{g}) = \bigvee_{\gamma \in \Gamma} \left(\bigvee_{\rho \tau = \gamma} f_\rho \cdot g_\tau \right) \gamma,$$

i.e., the Cauchy product of two sequences can be interpreted as the product of their formal series. It is important to notice that the Cauchy product of two sequences is well defined since the sequences take their values in the Boolean algebra $\mathbb{B}$.

If $a \in \mathbb{B}$ and $\underline{f} \in \Sigma(\Gamma, \mathbb{B})$, then $a \wedge \underline{f}$ denotes the sequence $(a \wedge \underline{f})(\gamma) = a \wedge \underline{f}(\gamma)$. This is either the zero sequence if $a = \underline{0}$ or it is the sequence $\underline{f}$ if $a = 1$.

With regard to the automaticity of sequences in $\Sigma(\Gamma, \mathbb{B})$ we have the following first result.

Lemma 4.1.2. *If $\underline{f}, \underline{g} \in \Sigma(\Gamma, \mathbb{B})$ are both H-automatic, then the sum $\underline{f} \vee \underline{g}$ and the $\wedge$-product $\underline{f} \wedge \underline{g}$ are both H-automatic.*

Proof. One easily shows that

$$\ker_{V,H}(\underline{f} \vee \underline{g}) \subset \{\underline{i} \vee \underline{j} \mid \underline{i} \in \ker_{V,H}(\underline{f}), \ \underline{j} \in \ker_{V,H}(\underline{g})\}$$

and

$$\ker_{V,H}(\underline{f} \wedge \underline{g}) \subset \{\underline{i} \wedge \underline{j} \mid \underline{i} \in \ker_{V,H}(\underline{f}), \ \underline{j} \in \ker_{V,H}(\underline{g})\}.$$

Since $\underline{f}$ and $\underline{g}$ are H-automatic, the sets on the right-hand side of the above inclusions are finite. $\square$

With this notation we can state several properties of H-automatic subsets.

Lemma 4.1.3. *Let $M, N \subset \Gamma$ be H-automatic subsets. Then the following holds:*

1. *The union $M \cup N$ is H-automatic.*

2. *The intersection $M \cap N$ is H-automatic.*

3. *The complement $\Gamma \setminus M = M^c$ is H-automatic.*

4. *The symmetric difference $M \triangle N = (M \cap N^c) \cup (M^c \cap N)$ is H-automatic.*

Proof. The set theoretic operations translate into operations with characteristic sequences having values in the Boolean algebra $\mathbb{B}$. E.g., $\chi_{M \cup N} = \chi_M \vee \chi_N$, $\chi_{M \cap N} = \chi_M \wedge \chi_N$, and $\chi_{M^c} = (\chi_M)^-$.

Then 1, 2, and 3 are immediate from Lemma 4.1.2. Assertion 4 follows from

$$\chi_{M \triangle N} = (\chi_M \wedge \chi_{\overline{N}}) \vee (\chi_{\overline{M}} \wedge \chi_N)$$

and Lemma 4.1.2. $\square$

The next lemma justifies that a study of automatic subsets of Γ is indeed a study of automatic sequences in $\Sigma(\Gamma, \mathcal{A})$. Before stating the lemma we define a 'multiplication' $* : \overline{\mathcal{A}} \times \mathbb{B} \to \overline{\mathcal{A}}$ by

$$a * b = \begin{cases} a & \text{if } b = 1 \\ \varnothing & \text{if } b = 0. \end{cases}$$

Let a be in $\mathcal{A}$ and let $\chi \in \Sigma(\Gamma, \mathbb{B})$, then $a * \chi$ is an element of $\Sigma(\Gamma, \overline{\mathcal{A}})$. It is defined by

$$(a * \chi)(\gamma) - a * \chi(\gamma).$$

Furthermore, we have the obvious identity

$$\partial_v^H(a * \chi) = a * \partial_v^H(\chi).$$

For $a \in \mathcal{A}$ we define the maps $\delta_a : \mathcal{A} \to \mathbb{B}$ by

$$\delta_a(b) = \begin{cases} 1 & \text{if } a = b \\ 0 & \text{otherwise.} \end{cases}$$

Lemma 4.1.4. *Let $\underline{f} \in \Sigma(\Gamma, \mathcal{A})$. The sequence $\underline{f}$ is H-automatic if and only if the sets*

$$M_a(\underline{f}) = \{\gamma \mid \underline{f}(\gamma) = a\}$$

are H-automatic subsets of Γ for all $a \in \mathcal{A}$.

Proof. For $a \in \mathcal{A}$ we define $\chi_a \in \Sigma(\Gamma, \mathbb{B})$ to be the characteristic sequence of the set $M_a(\underline{f})$. Then the sequence $\underline{f}$ can be written as

$$\underline{f} = \bigoplus_{u \in \mathcal{A}} a * \chi_{M_a(\underline{f})}$$

and we have $\chi_{M_a(\underline{f})} = \hat{\delta}_a(\underline{f})$. Since $\hat{\delta}_a \circ \partial_v^H = \partial_v^H \circ \hat{\delta}_a$, the assertion follows.   □

Therefore a study of automatic sequences is basically a study of automatic subsets of Γ. All of our previous theorems for automatic sequences apply to automatic subsets $M \subset \Gamma$. I.e., if V_c is a complete digit set for H, then there exists a finite automaton that generates the subset. In that case we can – in a more poetic language – say that the automaton answers to the question whether a given γ is an element of M with 'yes' or 'no'.

As we have seen, the usual set theoretic operations with automatic subsets do not affect the automaticity property. Besides these set theoretic operations there are operations with subsets of Γ which are induced by the group structure of Γ. The product MN of two subsets $M, N \subset \Gamma$ is defined by

$$MN = \{\gamma\rho \mid \gamma \in M, \ \rho \in N\}.$$

In terms of characteristic functions we have the following result

Lemma 4.1.5. *Let M, N be subsets of Γ. Then the characteristic sequence $\chi_{MN} \in \Sigma(\Gamma, \mathbb{B})$ of MN is the Cauchy product of χ_M with χ_N, i.e.,*

$$\chi_{MN} = \chi_M \cdot \chi_N.$$

Proof. Let $\gamma = \rho\tau$, where $\rho \in M$ and $\tau \in N$, then

$$(\chi_M \cdot \chi_N)(\gamma) = \bigvee_{\rho'\tau'=\gamma} \chi_M(\rho') \wedge \chi_N(\tau')$$

is equal to 1. On the other hand, if $(\chi_M \wedge \chi_N)(\gamma) = 1$, then there exists at least one pair (ρ', τ') such that $\rho' \cdot \tau' = \gamma$ and $\rho' \in M, \tau' \in N$.   □

Thus the question on the automaticity of MN translates into a question on the automaticity of the Cauchy product of two automatic sequences.

The definition of the products of sets allows us to define an embedding $\iota : \Gamma \to \Sigma(\Gamma, \mathbb{B})$ by setting

$$\iota(\gamma) = \chi_{\{\gamma\}}.$$

Then we have $\iota(\gamma\rho) = \iota(\gamma) \cdot \iota(\rho)$. If we interpret $\iota(\gamma)$ as a formal series, we can simply write $\iota(\gamma) = \gamma$. From now on we do not distinguish between $\iota(\gamma)$ and γ. It will always be clear from the context which interpretation of γ has to be chosen. The

Cauchy product enables us to give a new meaning to the left translation $(T_\tau)_*$ and right translation $(R_\tau)_*$. By their definition, see Equations (1.3), (1.4), we have

$$(T_\tau)_*(\underline{f}) = \tau \cdot \underline{f} \quad \text{and} \quad (R_\tau)_*(\underline{f}) = \underline{f} \cdot \tau$$

for all $\underline{f} \in \Sigma(\Gamma, \mathbb{B})$ and all $\tau \in \Gamma$, where we interpret τ as an element of $\Sigma(\Gamma, \mathbb{B})$. Due to Theorem 3.2.5, the left translations preserve the automaticity while the right translations do not necessarily preserve the automaticity, as shown by the example on p. 78. Since the Cauchy product $\chi_M \cdot \chi_N$ involves the left translations as well as the right translations, the question on the automaticity of $\chi_M \cdot \chi_N$ is closely related to the automaticity properties of left and right translations.

If Γ is a commutative group, then Γ is isomorphic to $\mathbb{Z}^n$, see Lemma 2.1.3, and the question on the automaticity of the product of two automatic sequences is answered in the next theorem. Before we state and prove this theorem we need a preparatory lemma. It provides us with a product formula for decimations and does only apply to the situation where Γ is commutative.

Lemma 4.1.6. *Let* $\underline{f}, \underline{g} \in \Sigma(\mathbb{Z}^N, \mathbb{B})$*, then*

$$\partial_v^H(\underline{f} \cdot \underline{g}) = \bigvee_{\substack{u,w \in V \\ \zeta(uw)=v}} \kappa(uw) \cdot \partial_u^H(\underline{f}) \cdot \partial_w^H(\underline{g}),$$

where κ *and* ζ *are the image-part- and remainder-map associated with* H *and a residue set* V *for* H.

Proof. Due to Lemma 2.2.16, we can write

$$\underline{f} = \bigvee_{v \in V} v \cdot H_*(\partial_v^H(\underline{f}))$$

$$\underline{g} = \bigvee_{v \in V} v \cdot H_*(\partial_v^H(\underline{g})).$$

Therefore the Cauchy product is given by

$$\underline{f} \cdot \underline{g} = \bigvee_{u,w \in V} uw \cdot H_*(\partial_u^H(\underline{f})) \cdot H_*(\partial_w^H(\underline{g})),$$

where we use the fact that the underlying group is commutative. If we write $uw = \zeta(uw)H(\kappa(uw))$ (again the commutativity is crucial), and if we use the fact that

$$H_*(\partial_u^H(\underline{f})) \cdot H_*(\partial_w^H(\underline{g})) = H_*(\partial_u^H(\underline{f}) \cdot \partial_w^H(\underline{g})).$$

then we can rewrite the product as

$$\underline{f} \cdot \underline{g} = \bigvee_{v \in V} v \cdot \bigvee_{\substack{u,w \in V \\ \zeta(uw)=v}} H(\kappa(uw)) \cdot H_*(\partial_u^H(\underline{f}) \cdot \partial_w^H(\underline{g})).$$

Observe that the summands can be written as

$$H\big(\kappa(uw)\big) \cdot H_*\big(\partial_u^H(\underline{f}) \cdot \partial_w^H(\underline{g})\big) = H_*\big(\kappa(uw) \cdot \partial_u^H(\underline{f}) \cdot \partial_w^H(\underline{f})\big)$$

and therefore the product takes the form

$$\underline{f} \cdot \underline{g} = \bigvee_{v \in V} v \cdot H_*\Bigg(\bigvee_{\substack{u,w \in V \\ \zeta(uw)=v}} \kappa(uw) \cdot \partial_u^H(\underline{f}) \cdot \partial_w^H(\underline{f}) \Bigg).$$

Thus, as a consequence of Lemma 2.2.16, $\partial_v^H(\underline{f} \cdot \underline{g})$ is of the desired form. $\qquad\square$

We are now prepared to prove the announced theorem on the automaticity of a product of automatic sets.

Theorem 4.1.7. *Let Γ be a commutative group and let $H : \Gamma \to \Gamma$ be an expanding endomorphism. If M and N are H-automatic subsets of Γ, then the product MN is H-automatic.*

Proof. Let $r = \max\{\|v\| \mid v \in V\}$ for a residue set for H and let C denote the expansion ratio of H. We define the set $K \subset \Sigma(\Gamma, \mathbb{B})$ by

$$K = \Bigg\{ \bigvee_{\rho, \underline{f}, \underline{g}} a_{\rho, \underline{f}, \underline{g}} \wedge (\rho \cdot \underline{f} \cdot \underline{g}) \Big| \ \|\rho\| \leq R, \ \underline{f} \in \ker_{V,H}(\chi_M),$$

$$\underline{g} \in \ker_{V,H}(\chi_N), a_{\rho, \underline{f}, \underline{g}} \in \mathbb{B} \Bigg\}.$$

where $R = \frac{5r}{C-1}$. Since M and N are automatic, K is a finite set. It remains to show that K is decimation invariant and contains $\chi_M \cdot \chi_N$. The latter assertion is certainly true. Therefore it remains to establish the decimation invariance of K. It is obvious that it suffices to prove that the decimation of a summand

$$a \wedge (\rho \cdot \underline{f} \cdot \underline{g}) = (\rho \cdot \underline{g} \cdot \underline{g})$$

is also an element of K. For $a = 0$ there is nothing to show. We therefore assume that $a = 1$. By Lemma 4.1.6, we can write

$$\rho \cdot \underline{f} \cdot \underline{g} = \rho \cdot \Bigg(\bigvee_{v \in V} v \cdot H_*\Bigg(\bigvee_{\substack{u,w \in V \\ \zeta(uw)=v}} \kappa(uw) \cdot \partial_u^H(\underline{f}) \cdot \partial_w^H(\underline{f}) \Bigg) \Bigg).$$

Since $v\rho = \zeta(v\rho) H(\kappa(v\rho))$, we obtain

$$\rho \cdot \underline{f} \cdot \underline{g} = \bigvee_{v \in V} \zeta(v\rho) H_*\Bigg(\bigvee_{\substack{u,w \in V \\ \zeta(uw)=v}} \kappa(v\rho)\kappa(uw)\partial_u^H(\underline{f}) \cdot \partial_w^H(\underline{g}) \Bigg).$$

Thus the decimations of $\rho \cdot \underline{f} \cdot \underline{g}$ are of the form

$$\bigvee_{\substack{u,w \in V \\ \zeta(uw)=v}} \kappa(v\rho)\kappa(uw)\partial_u^H(\underline{f}) \cdot \partial_w^H(\underline{g}).$$

It remains to estimate the norm of $\kappa(v\rho)\kappa(uw)$. We have

$$\|\kappa(v\rho)\kappa(uw)\| \leq \|\kappa(v\rho)\| + \|\kappa(uw)\| \leq \frac{\|v\| + \|\rho\| + \|u\| + \|v\| + 2r}{C}$$
$$\leq \frac{5r + \|\rho\|}{C}$$
$$\leq \frac{5r}{C-1},$$

where the last inequality is due to our choice of ρ. Thus the decimation of a summand is the sum of summands of the given form. Therefore the set K is decimation invariant which yields the H-automaticity of χ_{MN}. $\qquad\square$

The following lemma deals with a proper mixing of H-automatic subsets.

Lemma 4.1.8. *Let W be a residue set of G, where G satisfies the conditions of Lemma 3.2.11, and let $M_w \subset \Gamma$ with $w \in W$ be H-automatic subsets. The set*

$$M = \bigcup_{w \in W} wG(M_w),$$

where $wG(M_w) = \{wG(m) \mid m \in M_w\}$, is H-automatic.

Proof. The proof follows from the fact that

$$\chi_M = \bigoplus_{w \in W} (T_w \circ H)_*(\chi_{M_w})$$

and Theorems 3.2.5 and 3.2.12. $\qquad\square$

As a next step we discuss automatic sets of direct product of groups. We suppose that Γ_1 and Γ_2 are two finitely generated groups and $H_1 : \Gamma_1 \to \Gamma_1$ and $H_2 : \Gamma_2 \to \Gamma_2$ are expanding endomorphisms, respectively. The direct product $\Gamma = \Gamma_1 \times \Gamma_2$ is a group and the endomorphism $H = (H_1 \times H_2) : \Gamma \to \Gamma$ defined by $H(\gamma_1, \gamma_2) = (H_1(\gamma_1), H_2(\gamma_2))$ is expanding with respect to the discrete norm $\|(\gamma_1, \gamma_2)\| = \|\gamma_1\|_1 + \|\gamma_2\|_2$, where $\| \ \|_1$ and $\| \ \|_2$ are the discrete norms on Γ_1 and Γ_2. If V_1 and V_2 are residue sets for H_1 and H_2, then $V = V_1 \times V_2$ is a residue set for H.

Thanks to our general setting, all results stated so far carry over directly to the case where $\Gamma = \Gamma_1 \times \Gamma_2$ and $H = (H_1 \times H_2)$ is an expanding endomorphism.

Lemma 4.1.9. *Let* $\Gamma = \Gamma_1 \times \Gamma_2$ *and* $H = (H_1 \times H_2)$, *where* H_1, H_2 *are expanding endomorphisms of* Γ_1 *and* Γ_2. *If* $M_1 \subset \Gamma_1$ *and* $M_2 \subset \Gamma_2$ *are* H_1-*automatic and* H_2-*automatic, respectively, then the set* $M_1 \times M_2 \subset \Gamma$ *is* H-*automatic.*

Proof. The proof is almost trivial, all we have to do is to construct the characteristic sequence of the set $M_1 \times M_2$. It is defined by

$$\chi_{M_1 \times M_2}(\gamma_1, \gamma_2) = (\chi_{M_1} \odot \chi_{M_2})(\gamma_1, \gamma_2) := \chi_{M_1}(\gamma_1) \wedge \chi_{M_2}(\gamma_2),$$

i.e., $\chi_{M_1 \times M_2} = \chi_{M_1} \odot \chi_{M_2}$. Then the $(V_1 \times V_2, H)$-kernel of $\chi_{M_1 \times M_2}$ is contained in the finite set

$$K = \{\underline{f} \odot \underline{g} \mid \underline{f} \in \ker_{V_1, H_1}(\chi_{M_1}), \underline{g} \in \ker_{V_2, H_2}(\chi_{M_2})\}$$

and K is invariant under (v_1, v_2)-decimations. $\square$

Corollary 4.1.10. *Let* $\epsilon : \Sigma(\Gamma_1, \mathbb{B}) \to \Sigma(\Gamma_1 \times \Gamma_2, \mathbb{B})$ *be defined by*

$$\epsilon(\underline{f})(\gamma_1, \gamma_2) = \underline{f}(\gamma_1).$$

If $\underline{f} \in \Sigma(\Gamma_1, \mathbb{B})$ *is* H_1-*automatic, then* $\epsilon(\underline{f})$ *is* $(H_1 \times H_2)$-*automatic.*

Proof. $\epsilon(\underline{f})$ is the characteristic sequence of the set $\text{supp}(\underline{f}) \times \Gamma_2$. Therefore the assertion follows from Lemma 4.1.9. $\square$

The following result allows a kind of reduction of variables.

Theorem 4.1.11. *Let* $\underline{f}, \underline{g} \in \Sigma(\Gamma_1 \times \Gamma_2, \mathbb{B})$ *be* $(H_1 \times H_2)$-*automatic. The sequence* $\underline{h} = \underline{f} \otimes \underline{g} \in \Sigma(\Gamma_1, \mathbb{B})$ *defined by*

$$\underline{h}(\gamma_1) = \bigvee_{\gamma_2 \in \Gamma_2} \underline{f}(\gamma_1, \gamma_2) \wedge \underline{g}(\gamma_1, \gamma_2)$$

for $\gamma_1 \in \Gamma_1$ *is* H_1-*automatic.*

Proof. Let V_i be a fixed residue set for H_i, $i = 1, 2$. By $\ker(\underline{f})$ and $\ker(\underline{g})$ we denote the (V_1, H_1)-, (V_2, H_2)-kernel of $\underline{f}$ and $\underline{g}$, respectively. We define the subset $K \subset \Sigma(\Gamma_1, \mathbb{B})$ by

$$K = \left\{ \bigvee_{\substack{\underline{f}' \in \ker(\underline{f}), \\ \underline{g}' \in \ker(\underline{g})}} a_{\underline{f}', \underline{g}'} \wedge (\underline{f}' \otimes \underline{g}') \mid a_{\underline{f}', \underline{g}'} \in \mathbb{B} \right\} \cup \{\underline{0}\},$$

The set K is finite and contains $\underline{f} \otimes \underline{g}$. It remains to prove that K is decimation invariant. To this end it is sufficient to consider a summand $\underline{h}' = \underline{f}' \otimes \underline{g}', \underline{f}' \in \ker(\underline{f})$,

$\underline{g}' \in \ker(g)$, and to prove that $\partial_v^{H_1}(\underline{f}' \otimes \underline{g}') \in K$ for all $v \in V_1$. To this end we compute

$$\partial_v^{H_1}(\underline{h}')(\gamma_1) = \bigvee_{\gamma_2 \in \Gamma_2} \underline{f}'(vH(\gamma_1), \gamma_2) \cdot \underline{g}'(vH(\gamma_1), \gamma_2)$$

$$= \bigvee_{\substack{w \in V_2 \\ \gamma_2 \in \Gamma_2}} \underline{f}'(vH(\gamma_1), wH(\gamma_2)) \cdot \underline{g}(vH(\gamma_1), wH(\gamma_2))$$

$$= \bigvee_{w \in V_2} \bigvee_{\gamma_2 \in \Gamma_2} \underline{f}'(vH(\gamma_1), wH(\gamma_2)) \cdot \underline{g}(vH(\gamma_1), wH(\gamma_2)).$$

Due to the definition of $\otimes$, the last expression can be written as

$$\bigvee_{w \in V_2} \partial_{(v,w)}^{(H_1 \times H_2)}(\underline{f}') \otimes \partial_{(v,w)}^{(H_1 \times H_2)}(\underline{g}').$$

This shows that $\partial_v^{H_1}(\underline{h}')$ is an element of K. Therefore the set K is invariant under $\partial_v^{H_1}$-decimations. This completes the proof. $\qquad \square$

As a consequence of the above proof we note

Corollary 4.1.12. *If $\underline{f}, \underline{g} \in \Sigma(\Gamma_1 \times \Gamma_2, \mathbb{B})$, then*

$$\partial_{v_0}^{H_1} \circ \cdots \circ \partial_{v_k}^{H_1}(\underline{f} \otimes \underline{g})$$

$$= \bigvee_{w_0, \ldots, w_k \in V_2} \left(\partial_{(v_0, w_0)}^{(H_1 \times H_2)} \circ \cdots \circ \partial_{(v_k, w_k)}^{(H_1 \times H_2)}(\underline{f}) \right) \otimes \left(\partial_{(v_0, w_0)}^{(H_1 \times H_2)} \circ \cdots \circ \partial_{(v_k, w_k)}^{(H_1 \times H_2)}(\underline{g}) \right)$$

holds for all $v_0, \ldots, v_k \in V_1$ and all $k \in \mathbb{N}$.

The next lemma deals with projections of automatic sets.

Lemma 4.1.13. *Let Γ and H be as in Lemma 4.1.9. If $M \subset \Gamma$ is H-automatic, then the projection*

$$p_1(M) = \{\gamma_1 \in \Gamma_1 \mid \text{there exists } \gamma_2 \in \Gamma_2 \text{ with } (\gamma_1, \gamma_2) \in M\}$$

is H_1-automatic.

Proof. The characteristic sequence $\chi_{p_1(M)}$ is given by

$$\chi_{p_1(M)}(\gamma_1) = \bigvee_{\gamma_2 \in \Gamma_2} \chi_M(\gamma_1, \gamma_2) = (\chi_M \otimes \chi_{\Gamma_1 \times \Gamma_2})(\gamma_1)$$

and Theorem 4.1.11 applies. $\qquad \square$

It is obvious that the projection of an $(H_1 \times H_2)$-automatic subset of $\Gamma_1 \times \Gamma_2$ on the second component also yields an H_2-automatic subset.

As an application of Theorem 3.2.6, we have the following result.

Corollary 4.1.14. *Let* $\Gamma^2 = \Gamma \times \Gamma$ *and let* H *be an expanding endomorphism of* Γ. *If* $M \subset \Gamma^2$ *is* $(H \times H)$*-automatic, then the set*

$$M^T = \{(\gamma, \rho) \mid (\rho, \gamma) \in M\}$$

is $(H \times H)$*-automatic.*

Proof. The map $G : \Gamma^2 \to \Gamma^2$ defined by $G(\gamma, \rho) = (\rho, \gamma)$ is a group endomorphism which commutes with $H \times H$. Therefore Theorem 3.2.6 applies.   □

Remark. An analogous result is true for $(H_1 \times H_2)$-automatic subsets $M \subset \Gamma_1 \times \Gamma_2$. If M is $(H_1 \times H_2)$-automatic, then the set

$$M^T = \{(\gamma_2, \gamma_1) \mid (\gamma_1, \gamma_2) \in M\} \subset \Gamma_2 \times \Gamma_1$$

is $(H_2 \times H_1)$-automatic.

4.2 Automatic maps

In this section, we introduce the concept of automatic maps. The main purpose is, besides the interest in its own right, to obtain a large class of maps which preserve the automaticity of a sequence.

At the end of the section we shall show that the study of automatic maps $G : \Gamma_1 \to \Gamma_2$ can be reduced to a study of automatic maps on $\mathbb{N}$.

Definition 4.2.1. Let Γ_1, Γ_2 be groups, and H_1 and H_2 be expanding endomorphisms of Γ_1 and Γ_2. A map $G : \Gamma_1 \to \Gamma_2$ is called $(H_1 \times H_2)$-*automatic* if the graph of the map, i.e., set

$$\mathrm{Gr}(G) = \{(\gamma_1, G(\gamma_1)) \mid \gamma_1 \in \Gamma_1\},$$

is an $(H_1 \times H_2)$-automatic subset of $\Gamma_1 \times \Gamma_2$.

Remarks.

1. $G : \Gamma_1 \to \Gamma_2$ is $(H_1 \times H_2)$-automatic if and only if $\chi_{\mathrm{Gr}(G)}$ is $(H_1 \times H_2)$-automatic.

2. If $G : \Gamma \to \Gamma$ is $(H \times H)$-automatic, we simply say that G is H-automatic provided there is no risk of confusion.

3. Constant maps are $(H_1 \times H_2)$-automatic for all expanding maps H_1, H_2.

4. If $M \subset \Gamma_1$ and $G : M \to \Gamma_2$ is a map, then G is $(H_1 \times H_2)$-automatic if the graph, i.e., the set

$$\mathrm{Gr}(G : M \to \Gamma_2) = \{(\gamma, G(\gamma)) \mid \gamma \in M\},$$

is an $(H_1 \times H_2)$-automatic subset of $\Gamma_1 \times \Gamma_2$. By Lemma 4.1.13, the set M has to be H_1-automatic.

The following lemma relates the maps discussed in Theorem 3.2.5 and Theorem 3.2.6 with automatic maps.

Lemma 4.2.2.

1. *The left translation $T_\tau : \Gamma \to \Gamma$ is a H-automatic map.*

2. *If $G : \Gamma_1 \to \Gamma_2$ is a group endomorphism such that $G \circ H_1 = H_2 \circ G$, then G is $(H_1 \times H_2)$-automatic.*

Proof. Both proofs follow similar arguments already encountered in the proofs of Theorems 3.2.5, 3.2.6.

1. We fix a residue set V for H. Set $r = \max\{\|v\| \mid v \in V\}$ and denote by C the expansion ratio of H.

The characteristic function of $\mathrm{Gr}(T_\tau)$ is given by

$$\chi_{\mathrm{Gr}(T_\tau)} = \bigvee_{\gamma \in \Gamma} (\gamma, \tau\gamma).$$

This can be written as

$$\chi_{\mathrm{Gr}(T_\tau)} = \bigvee_{\substack{v \in V \\ \gamma \in \Gamma}} (vH(\gamma), \tau v H(\gamma))$$

or as

$$\chi_{\mathrm{Gr}(T_\tau)} = \bigvee_{\substack{v \in V \\ \gamma \in \Gamma}} (vH(\gamma), \zeta(\tau v)H(\kappa(\tau v)\gamma)),$$

where $\zeta = \zeta_{H,V}$ and $\kappa = \kappa_{H,V}$ are the remainder- and image-part-map, respectively. The above formula gives

$$\partial_{(v,w)}^{(H \times H)}(\chi_{\mathrm{Gr}(T_\tau)}) = \bigvee_{\gamma \in \Gamma} (\gamma, \kappa(\tau v)\gamma)$$

for $w = \zeta(\tau v) \in V$ and for $w \in V$ such that $w \neq \zeta(\tau v)$ we have

$$\partial_{(v,w)}^{(H \times H)}(\chi_{\mathrm{Gr}(T_\tau)}) = \underline{0},$$

the characteristic sequence of the empty set.

The above reasoning leads us to consider the set

$$K = \left\{ \textstyle\sum_{\gamma \in \Gamma}(\gamma, \rho\gamma) \mid \|\rho\| \leq R \right\} \cup \{\underline{0}\},$$

where $R = \max\left\{\frac{2r}{C-1}, \|\rho\|\right\}$. K is a finite set and decimation invariant and contains $\chi_{\mathrm{Gr}(T_\tau)}$. This shows that χ_{Gr} has a finite $(H \times H)$-kernel.

2. The characteristic sequence of the graph of G is given by

$$\chi_{\mathrm{Gr}(G)} = \bigvee_{\gamma \in \Gamma_1} (\gamma, G(\gamma)).$$

Let V_2 be a residue set for H_2, let $r = \max\{\|v\|_2 \mid v \in V_2\}$ and $\tilde{r} = \max\{\|G(v)\|_2 \mid v \in V_1\}$, where V_1 is a residue set for H_1. If C denotes the expansion ration of H, then for $R = \frac{r+\tilde{r}}{C-1}$. The set

$$K = \left\{ \sum_{\vee \in \Gamma_1} (\gamma, \rho G(\gamma)) \mid \|\rho\| \le R \right\} \cup \{\underline{0}\}$$

is finite, contains $\chi_{\mathrm{Gr}(G)}$, and is decimation invariant. Therefore G is $(H_1 \times H_2)$-automatic. $\square$

The situation is different for right translations, see the example on page 78.

Examples.

1. Let $H : \Gamma \to \Gamma$ be H-automatic. The $(H \times H)$-kernel of $\chi_{\mathrm{Gr}(H)}$ contains the sequences $\chi_{\mathrm{Gr}(T_v \circ H)}$, where $v \in V$, and the sequence $\underline{0}$. To prove this assertion, consider

$$\chi_{\mathrm{Gr}(H)} = \sum_{\gamma \in \Gamma} (\gamma, H(\gamma)) = \sum_{\substack{v \in V \\ \gamma \in \Gamma}} (vH(\gamma), H(vH(\gamma))).$$

This gives $\partial^{H \times H}_{(v,w)}(\chi_{\mathrm{Gr}(H)}) = \underline{0}$ for all $v \in V$ and $w \in V \setminus \{e\}$. For the (v, e)-decimations with $v \in V$ one obtains

$$\partial^{H \times H}_{(v,e)}(\chi_{\mathrm{Gr}(H)}) = \sum_{\gamma \in \Gamma} (\gamma, vH(\gamma)) = \chi_{\mathrm{Gr}(T_v)}.$$

Similar arguments show that the decimations of $\chi_{\mathrm{Gr}(T_v)}$ are either $\underline{0}$ or of the form $\chi_{\mathrm{Gr}(T_w)}$ for a $w \in W$.

2. Let $\mathrm{id}_\Gamma : \Gamma \to \Gamma$ be the identity. Then id_Γ is $(H \times H)$-automatic, grace to 2. of Lemma 4.2.2. But (!) id_Γ is not $(H \times H^{\circ 2})$-automatic. This follows from

$$\left(\partial^{H \times H^{\circ 2}}_{(e,e)}\right)^{\circ N}(\chi_{\mathrm{id}_\Gamma}) = \sum_{\gamma \in \Gamma} (H^{\circ N}(\gamma), \gamma)$$

for $N \in \mathbb{N}$.

3. Let $\Gamma = \mathbb{Z}$ and let $G(x^j) = x^{kj+l}$ where $k, l \in \mathbb{Z}$, then G is H-automatic for every $H(x^j) = x^{pj}$ and $|p| \ge 2$.

4. Let $\Gamma = \mathbb{N}$ and let $H(x^j) = x^{2j}$, then $G(x^j) = x^{j^2}$ is not H-automatic. This is a special case of a general theorem which we shall discuss later, see Example 1, p. 111.

5. Let $\Gamma = \mathbb{N}$ and $H_1(x^j) = x^{2j}$, $H_2(x^j) = x^{3j}$, furthermore let $V_1 = \{x^0, x^1\}$ and $V_2 = \{x^0, x^1, x^2\}$. The map $G : \mathbb{N} \to \mathbb{N}$ is defined in the following way: If $n = a_0 + a_1 2^1 + \ldots a_k 2^k$ is the binary expansion of n, then $G(n) = a_0 + a_1 3^1 + \cdots + a_k 3^k$ is the value of G at n. The map G is an example of an $(H_1 \times H_2)$-automatic map.

We continue our discussion of automatic maps by considering allowable operations with automatic maps.

Lemma 4.2.3. *If $G : \Gamma_1 \to \Gamma_2$ is a bijective map which is $(H_1 \times H_2)$-automatic, then the inverse map $G^{-1} : \Gamma_2 \to \Gamma_1$ is an $(H_2 \times H_1)$-automatic map.*

Proof. The assertion follows immediately from $\mathrm{Gr}(G^{-1}) = \mathrm{Gr}(G)^T$ and Corollary 4.1.14. □

The next important question is whether compositions of automatic maps are automatic. The question is answered by the next theorem.

Theorem 4.2.4. *Let Γ_i, $H_i : \Gamma_i \to \Gamma_i$, $i = 1, 2, 3$, be groups and expanding maps as always. If $F : \Gamma_1 \to \Gamma_2$ is $(H_1 \times H_2)$-automatic and if $G : \Gamma_2 \to \Gamma_3$ is $(H_2 \times H_3)$-automatic, then the composition $G \circ F : \Gamma_1 \to \Gamma_3$ is $(H_1 \times H_3)$-automatic.*

Proof. We define the embeddings

$$\epsilon_1 : \Sigma(\Gamma_1 \times \Gamma_2, \mathbb{B}) \to \Sigma(\Gamma_1 \times \Gamma_2 \times \Gamma_3, \mathbb{B}), \quad \epsilon_1(\underline{f})(\gamma_1, \gamma_2, \gamma_3) = \underline{f}(\gamma_1, \gamma_2),$$

$$\epsilon_2 : \Sigma(\Gamma_2 \times \Gamma_3, \mathbb{B}) \to \Sigma(\Gamma_1 \times \Gamma_2 \times \Gamma_3, \mathbb{B}), \quad \epsilon_1(\underline{f})(\gamma_1, \gamma_2, \gamma_3) = \underline{f}(\gamma_2, \gamma_3).$$

Due to Corollary 4.1.10, both embeddings preserve automaticity, e.g., if $f \in \Sigma(\Gamma_2 \times \Gamma_3, \mathbb{B})$ is $(H_2 \times H_3)$-automatic, then $\epsilon_2(\underline{f})$ is $(H_1 \times H_2 \times H_3)$-automatic. We define the projection

$$p_{1,3} : \Sigma(\Gamma_1 \times \Gamma_2 \times \Gamma_3, \mathbb{B}) \to \Sigma(\Gamma_1 \times \Gamma_3, \mathbb{B})$$

by setting

$$p_{1,3}(\underline{f})(\gamma_1, \gamma_3) = \bigvee_{\gamma_2 \in \Gamma_2} \underline{f}(\gamma_1, \gamma_2, \gamma_3).$$

In terms of subsets, $p_{1,3}$ is the usual projection of a subset $M \in \Gamma_1 \times \Gamma_2 \times \Gamma_3$ on the first and third coordinate. By Lemma 4.1.13, $p_{1,3}$ preserves automaticity. If $\chi_{\mathrm{Gr}(F)} \in \Sigma(\Gamma_1 \times \Gamma_2, \mathbb{B})$ and $\chi_{\mathrm{Gr}(G)} \in \Sigma(\Gamma_2 \times \Gamma_3, \mathbb{B})$ denote the characteristic sequences of $\mathrm{Gr}(F)$ and $\mathrm{Gr}(G)$, respectively, then the sequence

$$\epsilon_2(\chi_{\mathrm{Gr}(G)}) \odot \epsilon_1(\chi_{\mathrm{Gr}(F)}) \in \Sigma(\Gamma_1 \times \Gamma_2 \times \Gamma_3, \mathbb{B})$$

defined by

$$\epsilon_2(\chi_{\mathrm{Gr}(G)}) \odot \epsilon_1(\chi_{\mathrm{Gr}(F)})(\gamma_1, \gamma_2, \gamma_3) = \epsilon_2(\chi_{\mathrm{Gr}(G)})(\gamma_2, \gamma_3) \wedge \epsilon_1(\chi_{\mathrm{Gr}(F)})(\gamma_1, \gamma_2)$$

is $(H_1 \times H_2 \times H_3)$-automatic, see Lemma 4.1.9. Therefore the sequence

$$p_{1,3}(\epsilon_2(\chi_{\mathrm{Gr}(G)}) \odot \epsilon_1(\chi_{\mathrm{Gr}(F)})) \in \Sigma(\Gamma_1 \times \Gamma_3, \mathbb{B})$$

is $(H_1 \times H_3)$-automatic. It remains to show that

$$p_{1,3}(\epsilon_2(\chi_{\mathrm{Gr}(G)}) \odot \epsilon_1(\chi_{\mathrm{Gr}(F)})) = \chi_{\mathrm{Gr}(G \circ F)}.$$

We have

$$p_{1,3}(\epsilon_2(\chi_{\mathrm{Gr}(G)}) \odot \epsilon_1(\chi_{\mathrm{Gr}(G)})(\gamma_1, \gamma_3)) = \bigvee_{\gamma_2 \in \Gamma_2} \chi_{\mathrm{Gr}(G)}(\gamma_2, \gamma_3) \cdot \chi_{\mathrm{Gr}(F)}(\gamma_1, \gamma_2)$$

and the sum is equal to 1 if and only if there exists a $\gamma_1 \in \Gamma_1$ such that $F(\gamma_1) = \gamma_2$ and such that $G(\gamma_2) = \gamma_3$. This shows that $p_{1,3}(\epsilon_2(\chi_{\mathrm{Gr}(G)}) \odot \epsilon_1(\chi_{\mathrm{Gr}(F)})) = \chi_{\mathrm{Gr}(G \circ F)}$ and it follows that $G \circ F$ is $(H_1 \times H_3)$-automatic.   □

Remark. If $F : M_1 \to M_2$ and $G : M_2 \to \Gamma_3$ are automatic maps on $M_1 \subset \Gamma_1$ and $M_2 \subset \Gamma_2$, respectively, then their composition $G \circ F : M_1 \to \Gamma_3$ is $(H_1 \times H_3)$-automatic on M_1.

If $F, G : \Gamma_1 \to \Gamma_2$ are $(H_1 \times H_2)$-automatic maps, then the group structure of Γ_2 also allows to consider the group product $F\,G$ of these maps. It is defined by $(F\,G)(\gamma) = F(\gamma)G(\gamma)$. As for the product of two sets $M, N \subset \Gamma$, the question whether the group product of two automatic maps is automatic is a very delicate one. Again the discrete Heisenberg group shows that the group product of two automatic maps need not be automatic. To this end we consider the identity id : $L_{\mathbb{Z}} \to L_{\mathbb{Z}}$ on the Heisenberg group and the map $G(\gamma) = a$, where a is one of the generating elements of the Heisenberg group (see Lemma 1.2.1). Both maps are $H_{2,2}$-automatic. However, the group product (id G) is the map (id G)$(\gamma) = \gamma\,a$, i.e., a right translation which is not $H_{2,2}$-automatic.

We tackle the problem of the automaticity of the product of two automatic functions in a more general setting.

Definition 4.2.5. Let $\mu : \Gamma \times \Gamma \to \Gamma$ be a map. If μ is $((H \times H) \times H)$-automatic, then μ is called an H-*automatic operation* on Γ.

Automatic operations allow us to construct new automatic functions from two automatic functions. If μ is an automatic operation on Γ_2 and if G, F are maps from Γ_1 to Γ_2, then $\mu(G, F)$ denotes a new map from Γ_1 to Γ_2. It is defined by

$$\mu(G, F)(\gamma) = \mu(G(\gamma), F(\gamma))$$

and called the μ-*product* of G and F.

Theorem 4.2.6. *Let* $\mu : \Gamma_2 \times \Gamma_2 \to \Gamma_2$ *be an* $((H_2 \times H_2) \times H_2)$-*automatic operation on* Γ_2 *and let* $G : \Gamma_1 \to \Gamma_2$, $F : \Gamma_1 \to \Gamma_2$ *be* $(H_1 \times H_2)$-*automatic maps. Then the* μ-*product,* $\mu(G, F) : \Gamma_1 \to \Gamma_2$, *of* G *and* F *is* $(H_1 \times H_2)$-*automatic.*

Proof. From the automaticity of G and F it follows immediately that the map $\iota : \Gamma_1 \to \Gamma_2 \times \Gamma_2$ defined by

$$\iota(\gamma) = (G(\gamma), F(\gamma))$$

is $(H_1 \times (H_2 \times H_2))$-automatic. Due to Theorem 4.2.4, the composition of μ with ι, i.e., $\mu \circ \iota(\gamma) = \mu(G(\gamma), F(\gamma))$, is $(H_1 \times H_2)$-automatic. □

Remarks.

1. If $\mu_1, \mu_2 : \Gamma_2^2 \to \Gamma_2$, are H_2-automatic operations on Γ_2, then

$$\mu_1(F_1, \mu_2(F_2, F_3))$$

   is $(H_1 \times H_2)$-automatic for $(H_1 \times H_2)$-automatic maps $F_i : \Gamma_1 \to \Gamma_2, i = 1, 2, 3$.

2. It is straightforward to generalize the above result to operations with more than two variables $\mu : \Gamma^k \to \Gamma$.

The notion of automatic operations allows us to answer the question whether products of automatic functions are again automatic.

Corollary 4.2.7. *If the product* $\mu(\gamma, \rho) = \gamma\rho$ *is an* H_2-*automatic operation on* Γ_2 *and if* G, F *are* $(H_1 \times H_2)$-*automatic maps from* Γ_1 *to* Γ_2, *then the product* GF *is an* $(H_1 \times H_2)$-*automatic map.*

Note that for the Heisenberg group the multiplication is not an $H_{2 \times 2}$-automatic operation.

Lemma 4.2.8. *If* Γ *is a commutative group, then the product* $\mu(\gamma, \rho) = \gamma\rho$ *is an* H-*automatic operation on* Γ.

Proof. We denote by $\chi \in \Sigma(\Gamma \times \Gamma \times \Gamma, \mathbb{B})$ the characteristic sequence of the set

$$\{(\gamma, \rho, \gamma\rho) \mid \gamma, \rho \in \Gamma\},$$

which is the graph of the multiplication. It remains to show that χ is $(H \times H \times H)$-automatic. Let V be a residue set for H and let $r = \max\{\|v\| \mid v \in V\}$. Let $R = \frac{2r}{C-1}$, where $C > 1$ is the expansion ration of H. Then the set

$$K = \left\{ \underline{f}_\tau = \bigvee_{\gamma, \rho \in \Gamma} (\gamma, \rho, \tau\gamma\rho) \mid \|\tau\| \leq R \right\} \cup \{\underline{0}\}$$

is a finite subset of $\Sigma(\Gamma \times \Gamma \times \Gamma, \mathbb{B})$ and contains χ. As usual, to prove the automaticity of χ it suffices to show that K is invariant under decimations. To this end, we observe that, due to the commutativity of Γ,

$$\tau v H(\gamma) w H(\rho) = \tau v w H(\gamma \rho) = \zeta(\tau v w) H(\kappa(\tau v w) \gamma \rho),$$

where $v, w \in V$ and ζ and κ are the remainder- and image-part-maps w.r.t. V and H. If $\underline{f}_\tau$ is an element of K, then we have

$$\partial^{(H \times H \times H)}_{(v, w, \zeta(\tau v w))}(\underline{f}_\tau) = \bigvee_{\gamma, \rho \in \Gamma} (\gamma, \rho, \kappa(\tau v w) \gamma \rho)$$

and for $u \neq \zeta(\tau v w)$ we have

$$\partial^{(H \times H \times H)}_{(v, w, u)}(\underline{f}_\tau) = 0.$$

Due to the choice of τ one has $\|\kappa(\tau v w)\| \leq R$. This proves the decimation invariance of K.   □

So far we have discussed properties of $(H_1 \times H_2)$-automatic maps with respect to operations, i.e, composition, μ-product, on the set of $(H_1 \times H_2)$-automatic maps. The true relevance of H-automatic maps is revealed by the following results.

Theorem 4.2.9. *Let* $G : \Gamma_1 \to \Gamma_2$ *be* $(H_1 \times H_2)$-*automatic. If* $\underline{f} \in \Sigma(\Gamma_2, \mathbb{B})$ *is* H_2-*automatic, then* $G^*(\underline{f}) \in \Sigma(\Gamma_1, \mathbb{B})$ *is* H_1-*automatic.*

Proof. We consider the embedding, see Corollary 4.1.10, $\epsilon : \Sigma(\Gamma_2, \mathbb{B}) \to \Sigma(\Gamma_1 \times \Gamma_2, \mathbb{B})$ by setting

$$\epsilon(\underline{f})(\gamma_1, \gamma_2) = \underline{f}(\gamma_2).$$

Note that $\epsilon(\underline{f}) = \chi_{\Gamma_1 \times \mathrm{supp}\, \underline{f}}$. This gives the following equation for all $\gamma_1 \in \Gamma_1$:

$$G^*(\underline{f})(\gamma_1) = \bigvee_{\gamma_2 \in \Gamma_2} \chi_{\mathrm{Gr}(G)}(\gamma_1, \gamma_2) \cdot \epsilon(\underline{f})(\gamma_1, \gamma_2).$$

Therefore $G^*(\underline{f})$ is H_1-automatic, due to Corollary 4.1.10 and Theorem 4.1.11.   □

Remark. If $G : \Gamma \to \Gamma$ is bijective and H-automatic, then $G^*(\underline{f})$ is H-automatic if $\underline{f}$ is an H-automatic sequence. Thus, loosely speaking, an automatic rearrangement of an automatic sequence preserves the automaticity.

Theorem 4.2.10. *Let* $G : \Gamma_1 \to \Gamma_2$ *be an* $(H_1 \times H_2)$-*automatic map. If* $M \subset \Gamma_1$ *is* H_1-*automatic, then* $G(M)$ *is* H_2-*automatic.*

Proof. The assertion follows from the identity

$$\chi_{G(M)} = \chi_{\mathrm{Gr}(G)} \otimes \chi_{M \times \Gamma_2}$$

and the automaticity of both $\chi_{\mathrm{Gr}(G)}$ and $\chi_{M \times \Gamma_2}$ (see Lemma 4.1.9) in combination with Theorem 4.1.11. □

Finally, we state a result for preimages.

Theorem 4.2.11. *Let* $G : \Gamma_1 \to \Gamma_2$ *be an* $(H_1 \times H_2)$*-automatic map. If* $M \subset \Gamma_2$ *is* H_2*-automatic, then the preimage*

$$G^{-1}(M) = \{\gamma_1 \mid \gamma_1 \in \Gamma_1, \, G(\gamma_1) \in M\}$$

of M *is* H_1*-automatic.*

Proof. The proof is a consequence of

$$G^{-1}(M) = p_1 \left((\Gamma_1 \times M) \cap \mathrm{Gr}(G)\right),$$

where $p_1 : \Gamma_1 \times \Gamma_2 \to \Gamma_1$ is the projection on the first component. □

4.3 Automata and automatic maps

This section is devoted to the study of the properties of finite $(H_1 \times H_2)$-automata that generate an automatic map. We begin with a modified kernel graph of an automatic sequence.

Definition 4.3.1. Let $f \in \Sigma(\Gamma, \mathcal{A})$ be an H-automatic sequence such that the cardinality of the range of f satisfies $|\mathcal{R}(f)| \geq 2$ and let $a \in \mathcal{R}(f)$. Let V be a residue set of H. The labeled, directed graph $\mathcal{G}^a_{V,H}(f) = (\ker^a_{V,H}, f, V, K)$, where

- the set of vertices is given by

$$\ker^a_{V,H}(f) = \{h \in \ker_{V,H}(f) \mid h \neq a\},$$

where a denotes the constant sequence with value a,

- the set of edges is given by

$$K = \{g \xrightarrow{v} h \mid g, h \in \ker^a_{V,H}(f) \text{ and } \partial^H_v(g) = h\},$$

is called the *reduced kernel graph* (w.r.t. a).

Remarks.

1. The condition $\left|\mathcal{R}(f)\right| \geq 2$ is introduced to avoid an empty reduced kernel graph.

2. The reduced kernel graph is obtained from the kernel graph by removing the constant sequence $\underline{a}$ from the vertices of the kernel graph and by deleting all arrows leading to or leaving from the constant sequence $\underline{a}$. If the sequence $\underline{a}$ is not an element of the kernel, then the reduced kernel graph and the kernel graph are the same.

3. If $f \in \Sigma(\Gamma, \mathbb{B})$ is an H-automatic sequence, then we agree that the *reduced kernel graph of f*, denoted by $\mathcal{G}_{V,H}^{*}$, always means the reduced kernel graph w.r.t. $\underline{0}$. The set $\overline{\ker}_{V,H}(f) \setminus \{\underline{0}\}$ is denoted by $\ker_{V,H}^{*}(f)$ and is called the reduced kernel.

Let $(\overline{\mathcal{B}}, \varnothing, \mathbb{B}, \omega, \{\alpha_{(v,w)} \mid (v, w) \in V_1 \times V_2\})$, where V_1, V_2 are complete digit sets for H_1, H_2, respectively, be an $(H_1 \times H_2)$-automaton. This automaton generates a subset of $\Gamma_1 \times \Gamma_2$. The question is, when is this generated subset a graph of a map $F : \Gamma_1 \to \Gamma_2$. In order to answer this question we have to consider the transition graph of the given automaton. We begin by introducing the notion of a path in the transition graph of an automaton.

To $(v_i, w_i)_{i=0}^{k-1} \in (V_1 \times V_2)^k$, where $k \in \mathbb{N}$, $k \neq 0$, we associate a path of length k in the transition graph of the automaton. The path starts at the vertex $\varnothing$ and moves along the edges labeled (v_i, w_i), $i = 0, \ldots, k-1$. A *path in the transition graph* is denoted by $\mathfrak{w} = (\varnothing; (v_i, w_i)_{i=0,\ldots,k-1})$. We agree to consider the vertex $\varnothing$ as a path of length 0 in the transition graph and denote it by $\mathfrak{w}_\varnothing = (\varnothing;)$. The set of paths in the transition graph is denoted by $\mathfrak{W}(V_1 \times V_2)$. We define two evaluation maps $\mathbf{e}_i : \mathfrak{W}(V_1, V_2) \to \Gamma_i$, $i = 1, 2$. The first evaluation map $\mathbf{e}_1 : \mathfrak{W}(V_1, V_2) \to \Gamma_1$ is defined by

$$\mathbf{e}_1(\mathfrak{w}) = \begin{cases} e_1 & \text{if } \mathfrak{w} = (\varnothing;) \\ v_0 H(v_1) \ldots H^{\circ k} & \text{if } \mathfrak{w} = (\varnothing; (v_i, w_i)_{i=0}^{k}). \end{cases}$$

The second evaluation map is defined in a similar way for the second components.

For a path $\mathfrak{w} \neq \mathfrak{w}_\varnothing$ in the transition graph of an automaton the *terminal vertex* of the path is denoted by $\mathbf{t}(\mathfrak{w}) \in \overline{\mathcal{B}}$. The terminal vertex of $\mathfrak{w}_\varnothing$ is equal to $\varnothing$. Note that

$$\mathbf{t}(\mathfrak{w}) = \alpha_{(v_{k-1}, w_{k-1})} \circ \cdots \circ \alpha_{(v_0, w_0)}(\varnothing).$$

The next lemma states a necessary criterion for an automaton to generate an automatic map.

Lemma 4.3.2. *Let $F : \Gamma_1 \to \Gamma_2$ be $(H_1 \times H_2)$-automatic. If $(\overline{\mathcal{B}}, \varnothing, \mathbb{B}, \omega, \{\alpha_{(v,w)} \mid (v, w) \in V_1 \times V_2\})$ is an $(H_1 \times H_2)$-automaton that generates $\mathrm{Gr}(F)$, then for every $\gamma_1 \in \Gamma_1 \setminus \{e_1\}$ there exists a path $\mathfrak{w} \in \mathfrak{W}(V_1 \times V_2)$ such that*

$$\gamma_1 = \mathbf{e}_1(\mathfrak{w}) \quad \text{and} \quad \omega(\mathbf{t}(\mathfrak{w})) = 1.$$

The proof is clear. The following definition characterizes automata that generate automatic maps.

Definition 4.3.3. Let $(\overline{\mathcal{B}}, \varnothing, \mathbb{B}, \omega, \{\alpha_{(v,w)} \mid (v, w) \in V_1 \times V_2\})$ be an $(H_1 \times H_2)$-automaton, where V_1, V_2 are complete digit sets for H_1 and H_2, respectively. Furthermore, for $\gamma_1 \in \Gamma_1$ let

$$\mathfrak{W}_{\gamma_1}(V_1 \times V_2) = \{\mathfrak{w} \mid \gamma_1 = \mathbf{e}_1(\mathfrak{w}), \ \mathfrak{w} \in \mathfrak{W}(V_1, V_2)\}.$$

The automaton has the *unique first component property* if the following holds: If $\mathfrak{w}_1$, $\mathfrak{w}_2 \in \mathfrak{W}_{\gamma_1}(V_1 \times V_2)$ such that $\omega(\mathbf{t}(\mathfrak{w}_1)) = \omega(\mathbf{t}(\mathfrak{w}_2)) = 1$, then $\mathbf{e}_2(\mathfrak{w}_1) = \mathbf{e}_2(\mathfrak{w}_2)$.

Remarks.

1. Let $(\overline{\mathcal{B}}, \varnothing, \mathbb{B}, \omega, \{\alpha_{(v,w)} \mid (v, w) \in V_1 \times V_2\})$ be an automaton with the unique first component property. If the set

$$\mathfrak{W}_{\gamma_1}^* = \{\mathfrak{w} \mid \mathfrak{w} \in \mathfrak{W}_{\gamma_1}(V_1 \times V_2) \text{ and } \omega(\mathbf{t}(\mathfrak{w})) = 1\}$$

   is not empty, then it contains a unique shortest path $\mathfrak{w}_{\gamma_1} \in \mathfrak{W}_{\gamma_1}(V_1 \times V_2)$ such that $\omega(\mathbf{t}(\mathfrak{w}_{\gamma_1})) = 1$. In case $\gamma_1 = e_1$ this shortest path maybe the empty path. This shortest path is denoted by $\mathfrak{w}_{\gamma_1}$.

   If $\mathfrak{w}_{\gamma_1} = (\varnothing, (v_i, w_i)_{i=0}^k)$ denotes the shortest path, then all other possible paths in $\mathfrak{W}_{\gamma_1}^*(V_1 \times V_2)$ are of the form $(\varnothing, (v_0, w_0) \ldots (v_k, w_k), (e_1, e_2), \ldots (e_1, e_2))$.

2. It is essential for the definition that both V_1 and V_2 are complete digit sets. One can extend the above definition to arbitrary residue sets taking the periodic points of the image-part-maps into account. The extended definition would involve several awkward technical details which, as it seems, do not create any deeper insight into the realm of automatic maps.

3. If $M \subset \Gamma_1 \times \Gamma_2$ is the graph of an automatic map $G : \Gamma_1 \to \Gamma_2$, then the kernel graph (w.r.t. complete digit sets) provides an example of a finite automaton with the unique first component property. Moreover, the reduced kernel graph has also the unique first component property.

4. It is straightforward to generalize the unique first component to H_1-automatic subsets of $M \subset \Gamma_1$. We say that the automaton $(\overline{\mathcal{B}}, \varnothing, \mathbb{B}, \omega, \{\alpha_{(v,w)} \mid (v, w) \in V_1 \times V_2\})$ has the unique first component w.r.t. M if for every $\gamma \in M$ there exists a unique path in the transition graph of the automaton.

The relevance of the above definition is justified by the following lemma.

Lemma 4.3.4. *Let $(\overline{\mathcal{B}}, \varnothing, \mathbb{B}, \omega, \{\alpha_{(v,w)} \mid (v, w) \in V_1 \times V_2\})$ be an $(H_1 \times H_2)$-automaton with the unique first component property. If the set $\mathfrak{W}_{\gamma_1}^* \neq \varnothing$ for all $\gamma_1 \in \Gamma_1$, then the set M generated by the automaton is the graph of an $(H_1 \times H_2)$-automatic map $F : \Gamma_1 \to \Gamma_2$.*

Proof. By the definition of an automaton with the unique first component property, the sets V_1 and V_2 are complete digit sets for H_1 and H_2, respectively. We define a map $F : \Gamma_1 \to \Gamma_2$ in the following way. Since $\mathfrak{W}^*_{\gamma_1} \neq \emptyset$ for all $\gamma_1 \in \Gamma_1$ there exists a unique shortest path $\mathfrak{w}_{\gamma_1}$ in $\mathfrak{W}^*_{\gamma_1}$. Define F as

$$F(\gamma_1) = \mathbf{e}_2(\mathfrak{w}_{\gamma_1}).$$

This defines a map F from Γ_1 to Γ_2 such that $\mathrm{Gr}(F) \subset M$.

It remains to show that the graph $\mathrm{Gr}(F)$ of F is equal to M. Therefore let us assume that there exists a pair $(\gamma_1, \gamma_2) \in M$ such that $F(\gamma_1) \neq \gamma_2$. Suppose that $(e_1, e_2) \in M$ and $(e_1, e_2) \notin \mathrm{Gr}(F)$, then the shortest path $\mathfrak{w}_{e_1}$ in $\mathfrak{W}^*_{e_1}$ satisfies $\mathbf{e}_1(\mathfrak{w}_{e_1}) = e_1$ and $\mathbf{e}_2(\mathfrak{w}_{e_1}) = e_2$. This shows that $F(e_1) = e_2$, a contradiction.

Now assume that $(\gamma_1, \gamma_2) \in M$, $(\gamma_1, \gamma_2) \notin \mathrm{Gr}(F)$ and $(\gamma_1, \gamma_2) \neq (e_1, e_2)$. If $(v_0, w_0)(H_1 \times H_2)(v_1, w_1) \ldots (H_1 \times H_2)^{\circ k}(v_k, w_k)$ denotes the unique representation of the pair (γ_1, γ_2) w.r.t. the complete digit set $V_1 \times V_2$, then it follows that

$$\omega(\alpha_{(v_k, w_k)} \circ \cdots \circ \alpha_{(v_0, w_0)}(\emptyset)) = 1.$$

If $v_k \neq e_1$, then the path $(\emptyset, (v_0, w_0), \ldots, (v_k, w_k))$ is the shortest path in $\mathfrak{W}^*_{\gamma_1}$, therefore $F(\gamma_1) = \gamma_2$. If $v_k = e_1$, then $w_k \neq e_2$ and therefore $(\emptyset, (v_0, w_0), \ldots, (v_k, w_k))$ is the shortest path in $\mathfrak{W}^*_{\gamma_1}$, i.e., $F(\gamma_1) = \gamma_2$. Therefore M is the graph of an automatic function.   □

Remark. If the automaton in Lemma 4.3.4 has the unique first component property w.r.t. an H_1-automatic subset $M \subset \Gamma_1$ and if $\mathfrak{W}^*_\gamma \neq \emptyset$ for all $\gamma \in M$, then the above arguments show that the automaton defines a function $G : M \to \Gamma_2$.

Examples.

1. Let $\Gamma = \langle x \rangle$ and $H(x^j) = x^{2j}$ and $V = \{x^0, x^1\}$. In Figure 4.1 the kernel graph of an automatic subset of $\mathbb{N}^2$ is shown. We have omitted all directed edges emanating from the vertex $\underline{0}$, the zero sequence. The automaton has the unique first component property and $\mathfrak{W}^*_{\gamma_1} \neq \emptyset$ for all $\gamma_1 \in \mathbb{N}$. Therefore the automaton represents an automatic function.

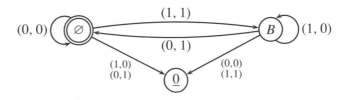

Figure 4.1. Transition graph with the unique first component property. Note that $\omega(\emptyset) = 1$, $\omega(B) = \omega(\underline{0}) = 0$.

2. Let $\Gamma = \langle x \rangle$ and $H(x^j) = x^{2j}$ and $V = \{x^0, x^1\}$. In Figure 4.2 a kernel graph without the unique first component property is depicted. The set $M \subset \mathbb{N}^2$ generated by this automaton is characterized by $M = \{(n, k) \mid \binom{k}{n} \text{ is odd}, n, k \in \mathbb{N}\}$. Again we have omitted the edges leaving from the zero sequence $\underline{0}$.

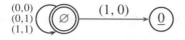

Figure 4.2. Transition graph without the unique first component property. Note that $\omega(\varnothing) = 1$, $\omega(\underline{0}) = 0$.

3. Let $\Gamma = \langle x \rangle$ and $H(x^j) = x^{2j}$ and $V = \{x^0, x^1\}$. Figure 4.3 shows the reduced kernel graph of an automatic subset $M \subset \mathbb{N}^2$. The output function is defined by $\omega(\varnothing) = \omega(C) = \omega(D) = 1$ and $\omega(B) = 0$. Note that all missing edges lead to the constant sequence $\underline{0}$. The automaton satisfies the requirements of Lemma 4.3.4 and therefore defines an automatic function $G : \mathbb{N} \to \mathbb{N}$. This function is in fact given by

$$G(n) = \begin{cases} 0 & \text{if } n = 0 \text{ or } n = 2^k \text{ for an } k \in \mathbb{N} \\ n & \text{otherwise.} \end{cases}$$

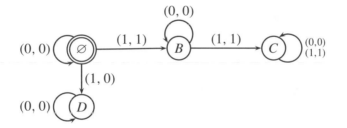

Figure 4.3. Reduced kernel graph of a subset $M \subset \mathbb{N}^2$ with unique first component property.

Although Examples 1 and 3 are examples for reduced kernel graphs with the unique first component property there exists a difference. The first example has exactly two edges emanating from each vertex. Moreover for each vertex g in the reduced kernel there exists a bijective map $v_g : \{0, 1\} \to \{0, 1\}$ such that the arrow $(v, v_g(v))$ ends in a vertex of the reduced kernel graph. This is not true for the third example.

Automata with this property will be studied next.

Definition 4.3.5. Let $M \subset \Gamma_1 \times \Gamma_2$ be an $(H_1 \times H_2)$-automatic set. The set M has the *transducer property* if there exist complete digit sets V_1 for H_1 and V_2 for H_2 such that for all $\underline{g} \in \ker^*_{(V_1 \times V_2),(H_1 \times H_2)}(\chi_M)$ there exists a map

$$v_g : V_1 \to V_2$$

such that

$$\partial_{(v,w)}^{(H_1 \times H_2)}(\underline{g}) \in \ker^*_{(V_1 \times V_2),(H \times H_2)}(\chi_{\mathrm{Gr}(G)})$$

if and only if $w = v_{\underline{g}}(v)$.

Remarks.

1. If $G : \Gamma_1 \to \Gamma_2$ is an $(H_1 \times H_2)$-automatic map, then we say that G has the transducer property if the set $\mathrm{Gr}(G)$ has the transducer property.

2. If $M \subset \Gamma_1 \times \Gamma_2$ is $(H_1 \times H_2)$-automatic and has the transducer property, then there exists a map $G : p_1(M) \to M$, where $p_1(M) = \{m_1 \mid (m_1, m_2) \in M\}$, such that $M = \{(\gamma, G(\gamma)) \mid \gamma \in p_1(M)\}$.

Note that we do not require that the maps v_g are bijective. At a first glance one would expect that the transducer property of a automatic function depends on the choice of the residue set. As a consequence of the following lemma we shall see that the transducer property is a property of the automatic function G.

Lemma 4.3.6. *Let $G : \Gamma_1 \to \Gamma_2$ be $(H_1 \times H_2)$-automatic. Let V_1 and V_2 be complete digit sets for H_1 and H_2, respectively. Then G has the transducer property w.r.t. $V_1 \times V_2$ if and only if every $\underline{g} \in \ker^*_{(V_1 \times V_2),(H_1 \times H_2)}(\chi_{\mathrm{Gr}(G)})$ is the graph of a function $G_{\underline{g}} : \Gamma_1 \to \Gamma_2$.*

Proof. Let G have the transducer property. Let $\ker^*_{V_1 \times V_2, H_1 \times H_2}(\chi_{\mathrm{Gr}(G)})$ be the reduced kernel of $\mathrm{Gr}(G)$. Then every kernel element $\underline{h}$ defines a subset $M_{\underline{h}} = \mathrm{supp}(\underline{h}) \subset \Gamma_1 \times \Gamma_2$. Since G has the transducer property, every subset $M_{\underline{h}}$ also has the transducer property. It is therefore sufficient to prove that the decimations $\partial_{(v,w)}^{(H_1 \times H_2)}(\chi_{\mathrm{Gr}(G)})$, $w = v_{\chi_{\mathrm{Gr}(G)}}(v)$, represent a map defined on Γ_1. We have

$$\chi_{\mathrm{Gr}(G)} = \bigvee_{\gamma \in \Gamma_1} (\gamma, G(\gamma))$$

$$= \bigvee_{\substack{v \in V \\ \gamma \in \Gamma}} (vH_1(\gamma), G(vH_1(\gamma)))$$

$$= \bigvee_{\substack{v \in V \\ \gamma \in \Gamma}} (vH_1(\gamma), \zeta(G(vH_1(\gamma)))H_2(\kappa(G(vH_1(\gamma))))),$$

where ζ and κ are the remainder- and image-part-maps w.r.t. V_2 and H_2. The transducer property now implies that

$$\zeta(G(vH_1(\gamma)))H_2(\kappa(G(vH_1(\gamma)))) = v_{\chi_{\mathrm{Gr}(G)}}(v) = w$$

holds for all $\gamma \in \Gamma_1$. Therefore we see that

$$\partial_{(v,w)}^{(H_1 \times H_2)}(\chi_{\mathrm{Gr}(G)}) = \bigvee_{\gamma \in \Gamma_1} (\gamma, \kappa(G(vH_1(\gamma))))$$

represents a function from Γ_1 to Γ_2. This shows that every set $M_{\underline{h}}$ is a graph of a map from Γ_1 to Γ_2.

Let us now assume that for every kernel element $\underline{g}$ the set $M_{\underline{g}}$ is a graph of a function from Γ_1 to Γ_2. To prove that G has the transducer property it suffices to show that there exists a map $\nu_{\underline{g}} : V_1 \to V_2$ such that $\partial^{H_1 \times H_2}_{(v, \nu_{(\underline{g})}(v))}(\underline{g})$ is an element of the reduced kernel of $\chi_{Gr(G)}$ and $\partial^{(H_1 \times H_2)}_{(v, w)}(\underline{g}) = \underline{0}$ for $w \neq \nu_{\underline{g}}(v)$. Let $v \in V_1$, then there exists a $w \in V_2$ such that $\partial^{(H_1 \times H_2)}_{(v, w)}(\underline{g}) \neq \chi_\emptyset$, otherwise $M_{\underline{g}}$ would not be a graph of a function. Then we have

$$\partial^{(H_1 \times H_2)}_{(v, w)}(\underline{g})(\gamma, \rho) = \underline{g}(v H_1(\gamma), w H_2(\rho)),$$

and for any γ there exists a unique ρ such that

$$\underline{g}(v H_1(\gamma), w H_2(\rho)) \neq 0,$$

since $\partial^{(H_1 \times H_2)}_{(v, w)}(\underline{g})$ represents a graph of a function. If there exists a $w' \neq w$ such that $\partial^{(H_1 \times H_2)}_{(v, w')}(\underline{g}) \neq \chi_\emptyset$, then there exists for any $\gamma \in \Gamma_1$ a unique ρ' such that

$$\partial^{(H_1 \times H_2)}_{v, w'}(\underline{g})(\gamma, \rho') = \underline{g}(v H(\gamma), w' H(\rho')) \neq 0.$$

Since $M_{\underline{g}}$ is a graph of function this is impossible unless $w' = w$ and $\rho' = \rho$. Thus for any $v \in V$ there exists a unique $w \in V$ such that $\partial^{(H_1 \times H_2)}_{(v, w)}(\underline{g})$ is an element of the reduced kernel of $\chi_{Gr(G)}$. In other words, G has the transducer property. $\square$

As a corollary to the above lemma we obtain that the transducer property of an automatic function is a genuine property of G.

Corollary 4.3.7. *Let $G : \Gamma_1 \to \Gamma_2$ be an $(H_1 \times H_2)$-automatic map such G has the transducer property w.r.t. $V_1 \times V_2$. Then G has the transducer property w.r.t. any residue set for $(H_1 \times H_2)$.*

Proof. If G has the transducer property w.r.t. $V_1 \times V_2$, then any kernel element represents a function. If $V_1' \times V_2'$ is another residue set for $(H_1 \times H_2)$, then, by Lemma 3.2.3, for every $\underline{g} \in \ker_{V_1' \times V_2'}(\chi_{Gr(G)})$ there exists a (γ_1, γ_2) such that

$$(T_{(\gamma_1, \gamma_2)})^*(\underline{g}) \in \ker_{V_1 \times V_2}(\chi_{Gr(G)}).$$

Therefore $\underline{g}$ represents a function defined on Γ_1. By the above lemma, the assertion follows. $\square$

Examples.

1. Let $\Gamma = \mathbb{N}$ and $H(x^j) = x^{jp}$ and let $k \in \mathbb{N}$. Then the function $F(x^j) = x^{kj}$ is H-automatic and has the transducer property. The function $F_m(x^j) = x^{kj+m}$ with $m \in \mathbb{N}, m \geq 1$, provides another class of H-automatic functions on $\mathbb{N}$ with the transducer property.

2. If $G : \Gamma \to \Gamma$ is a group endomorphism that commutes with H, then G has the transducer property.

3. Example 1, p. 104, shows an automatic function $G : \mathbb{N} \to \mathbb{N}$ that has the transducer property. It is not a function as simple as the examples given above. In fact, G is fairly complicated. Using the reduced kernel graph one can show that G satisfies the following inequalities:

$$n \leq G(n) \leq 3n.$$

These inequalities are best possible, since $G(2^n - 1) = 2^n + 1$ and $G(2^n) = 3 \cdot 2^n$ for all $n \geq 1$.

The last example shows that automatic functions may have a strange behaviour as far as their growth is concerned. We shall investigate this phenomenon in the next section.

By Theorem 4.2.10 we already know that $G(\Gamma_1)$ is H_2-automatic if the map $G : \Gamma_1 \to \Gamma_2$ is $(H_1 \times H_2)$-automatic. The image of Γ_1 under G is described by the sequence $\chi_{G(\Gamma_1)} \in \Sigma(\Gamma_2, \mathbb{B})$ which is also given by

$$\chi_{G(\Gamma_1)}(\gamma_2) = \bigvee_{\gamma_1 \in \Gamma_1} \chi_{\mathrm{Gr}(F)}(\gamma_1, \gamma_2).$$

The above formula allows us to compute the kernel graph of $\chi_{G(\Gamma_1)}$ using Corollary 4.1.12. An application of the above formula is particularly easy if the automatic map has the transducer property and the maps v_g are injective. Then the kernel graph of the image $G(\Gamma_1)$ is obtained by removing all first coordinates from the labels of the edges. The graph obtained in this way is the kernel graph of the characteristic sequence of the set $G(\Gamma_1)$.

Examples.

1. The automatic function defined in Figure 4.1 has the transducer property. Moreover, each map v_g is injective. Thus the automaton for the image $G(\mathbb{N})$ is given by removing the first coordinates from the labels. This shows that $G(\mathbb{N})$ is the support of the Thue–Morse sequence.

2. Figure 4.4 shows the reduced kernel graph of the graph of the function $G_0(n) = 3n$, $n \in \mathbb{N}$, w.r.t. $H_1(x^j) = H_2(x^j) = x^{2j}$ and $V_1 = V_2 = \{x^0, x^1\}$. The re-

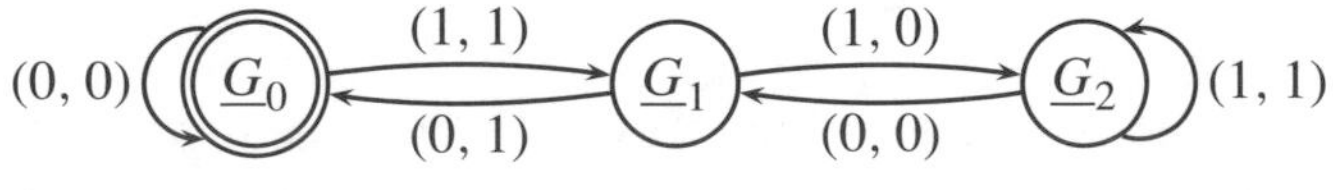

$$(0,0) \; \underline{G_0} \xrightarrow[\;(0,1)\;]{\;(1,1)\;} \underline{G_1} \xleftarrow[\;(0,0)\;]{\;(1,0)\;} \underline{G_2} \;(1,1)$$

Figure 4.4. Reduced kernel graph of the graph of the function $G_0(n) = 3n$.

duced kernel graph has the transducer property. Removing the first components
of the labels gives the kernel graph of $G_0(\mathbb{N}) = 3\mathbb{N}$. Moreover, the vertices
$G_{1,2}$ represent the functions $G_i(n) = 3n + i$, $i = 1, 2$.

The above example raises the question: If $M \subset \Gamma_2$ is an H_2-automatic subset,
does there exist an $(H_1 \times H_2)$-automatic map $G : \Gamma_1 \to \Gamma_2$ such that $G(\Gamma_1) = M$?
In its simplest form this question is related to the existence of an $(H_1 \times H_2)$-automatic
map G such that $G(\Gamma_1) = \Gamma_2$.

In order to answer these questions we introduce a kind of normalization. We shall
show that every $(H_1 \times H_2)$-automatic map can be realized in a precise sense by an
(x^p, x^q)-automatic map from $\mathbb{N}$ to $\mathbb{N}$. To this end we define the p-valuation map.

Definition 4.3.8. Let $H : \Gamma \to \Gamma$ be an expanding group endomorphism such that the
index of $H(\Gamma)$ is equal to $p \geq 2$ and let $V_c = \{e = v_0, v_1, \ldots, v_{p-1}\}$ be a complete
digit set for H. The map $\mathfrak{V}_p : \Gamma \to \mathbb{N}$ defined by

$$\mathfrak{V}_p(\gamma) = \mathfrak{V}_p(v_{i_0} H(v_{i_1}) \ldots H^{\circ k}(v_{i_k})) = \sum_{j=0}^{k} i_j p^j,$$

where $v_{i_0} H(v_{i_1}) \ldots H^{\circ k}(v_{i_k})$ denotes the unique representation of γ, is called a
p-valuation map (w.r.t. (H, V_c)).

Remarks.

1. A p-valuation map transforms the unique representation of $\gamma \in \Gamma$ w.r.t. the
   complete digit set V_c of H into the unique representation of a natural number
   $n \in \mathbb{N}$ w.r.t. the complete digit set $\{0, 1, \ldots, p - 1\}$ for the expanding map
   $n \mapsto pn$.

2. There exist several p-valuation maps depending on the numbering of the complete digit set V_c. In fact, every bijective map $v : V_c \setminus \{e\} \to \{1, \ldots, p - 1\}$
   defines a p-valuation map. However, from the point of view of automaticity all
   these p-valuations are equivalent.

3. A p-valuation map is completely determined by its values on V_c.

Lemma 4.3.9. *Let $\mathfrak{V}_p : \Gamma \to \mathbb{N}$ be a p-valuation map.*

1. *The map $\mathfrak{V}_p$ is bijective.*

2. *For all $\gamma \in \Gamma$ and all $v \in V_c$ the following holds:*

$$\mathfrak{V}_p(vH(\gamma)) = \mathfrak{V}_p(v) + p\mathfrak{V}_p(\gamma).$$

3. *If κ and ζ denote the remainder- and image-part-map w.r.t. V_c and H, respectively, then*

$$\mathfrak{V}_p(\kappa(\gamma)) = \frac{1}{p}\left(\mathfrak{V}_p(\gamma) - \mathfrak{V}_p(\zeta(\gamma))\right)$$

holds for all $\gamma \in \Gamma$.

The proof is an obvious consequence of the fact that V_c and $\{0, 1, \ldots, p-1\}$ are complete digit sets for H and $H_p(x^n) = x^{pn}$, respectively. The next lemma gives a first hint to the importance of the p-valuation map.

Lemma 4.3.10. *Let $H : \Gamma \to \Gamma$ such that the index of $H(\Gamma)$ is equal to p and let $V_c = \{v_0 = e, v_1, \ldots, v_{p-1}\}$ be a complete digit set for H. Let $\Gamma_2 = \mathbb{N}$, let $H_p : \mathbb{N} \to \mathbb{N}$ be defined by $H_p(n) = np$ and let $V_p = \{0, 1, \ldots, p-1\}$. Then a p-valuation map is $(H \times H_p)$-automatic.*

Proof. Figure 4.5 defines an automaton that generates the graph of $\mathfrak{V}_p : \Gamma \to \mathbb{N}$, see also Lemma 4.3.9. ☐

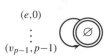

Figure 4.5. Reduced kernel graph of the graph of a p-valuation map, $\omega(\varnothing) = 1$.

As a corollary we note that as far as the H-automaticity of a set $M \subset \Gamma$ is concerned the actual choice of the p-valuation map is not important.

Corollary 4.3.11. *Let $\mathfrak{V}_p$ and $\mathfrak{V}_p'$ be p-valuation maps. The map $\mathfrak{V}_p^{-1} \circ \mathfrak{V}_p' : \Gamma \to \Gamma$ is bijective and $(H \times H)$-automatic.*

As a consequence of the above results we note

Theorem 4.3.12. *Let $H : \Gamma \to \Gamma$ be expanding such that the index of $H(\Gamma)$ is equal to p and let V_c be a complete digit set for H. A subset $M \subset \Gamma$ is H-automatic if and only if $\mathfrak{V}_p(M) \subset \mathbb{N}$ is H_p-automatic, where $H_p(x^j) = x^{pj}$.*

The notion of $(H_1 \times H_2)$-automatic maps $G : \Gamma_1 \to \Gamma_2$ can be transported to $\mathbb{N}$.

Lemma 4.3.13. *Let $G : \Gamma_1 \to \Gamma_2$ a map and let $\mathfrak{V}_p : \Gamma_1 \to \mathbb{N}$ and $\mathfrak{V}_q : \Gamma_2 \to \mathbb{N}$ be valuations (w.r.t. (H_1, V_1) and (H_2, V_2)). The map G is $(H_1 \times H_2)$-automatic if and only if*

$$\mathfrak{V}_q \circ G \circ \mathfrak{V}_p^{-1} : \mathbb{N} \to \mathbb{N}$$

is $(H_p \times H_q)$-automatic.

Let $M_i \subset \Gamma_i$ be (H_i, V_i)-automatic sets for $i = 1, 2$ and let $|V_1| = p$, $|V_2| = q$. Then $\mathfrak{V}_p(M_1)$ and $\mathfrak{V}_q(M_2)$ are both subsets of $\mathbb{N}$. It is therefore natural to ask under which conditions these sets are equal. This question can be rephrased as follows: Under which conditions is a set $M \subset \mathbb{N}$ both H_p-automatic and H_q-automatic? By Theorem 3.2.1 and Lemma 3.2.4, it follows that the H_p-automaticity of $M \subset \mathbb{N}$ implies the H_q-automaticity of M if there exist $\alpha, \beta \in \mathbb{N}$, $\alpha + \beta \neq 0$, such that

$$p^\alpha = q^\beta$$

holds. One calls p and q *multiplicatively independent* if the above equation has only the trivial solution $\alpha = \beta = 0$. In [56] a characterization of H_p- and H_q-automatic sets is given for multiplicatively independent p and q.

Theorem 4.3.14 (Cobham). *Let $p, q \in \mathbb{N}$, $p, q \geq 2$ be multiplicatively independent. A subset M of $\mathbb{N}$ is p-automatic and q-automatic if and only if there exist $n_0 \in \mathbb{N}$ and $d \in \mathbb{N} \setminus \{0\}$ such that*

$$\chi_M(n + d) = \chi_M(n)$$

holds for all $n \geq n_0$.

In other words, M is H_p- and H_q-automatic if and only if χ_M is ultimately periodic. Note that in this case M is automatic for all $p \geq 2$.

Examples.
1. The sets $M_p = \{p^n \mid n \in \mathbb{N}\}$, $p \geq 2$, are H_p-automatic and not H_q-automatic for multiplicatively independent p and q. Obviously, χ_{M_p} is not ultimately periodic and M_p is H_p-automatic. By Cobham's theorem M_p is not H_q-automatic.

2. The set $M = \{(n, n) \mid n \in \mathbb{N}\} \subset \mathbb{N}^2$ is (p, p)-automatic for all $p \geq 2$, $p \in \mathbb{N}$. For $p = 2$ and the complete digit set $V_1 = \{(0, 0), (1, 0), (0, 1), (1, 1)\}$ of $\Pi(l, m) = 2(l, m)$ we define a 4-valuation map $\mathfrak{V}_4$ as the continuation of

$$\mathfrak{V}_4(0, 0) = 0, \quad \mathfrak{V}_4(0, 1) = 1,$$
$$\mathfrak{V}_4(1, 0) = 2, \quad \mathfrak{V}_4(1, 1) = 3.$$

The set $\mathfrak{V}_4(M)$ is H_4-automatic. Figure 4.6 shows the reduced kernel graph of the characteristic sequence χ of $\mathfrak{V}_4(M)$.

$$0, 3\ \ (\!(\ \varnothing\ )\!)$$

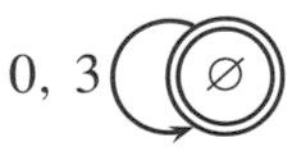

Figure 4.6. Reduced kernel graph of the characteristic sequence of $\mathfrak{V}_4(M)$.

From the reduced kernel graph one immediately sees that χ is not ultimately periodic. Therefore $\mathfrak{V}_4(M)$ is not H_q-automatic for all q multiplicatively independent from 4.

4.4   Automatic functions on $\mathbb{N}$

In this section we investigate properties of automatic maps defined on $\mathbb{N}$. We consider $(\mathbb{N}, +)$ as a semigroup. Then any expanding (w.r.t. the absolute value) endomorphism is given by $H_p : \mathbb{N} \to \mathbb{N}, n \mapsto pn$, where $p \geq 2, p \in \mathbb{N}$. Moreover, there exists only one complete digit set $V_p = \{0, 1, \ldots, p - 1\}$. However, all of our above results on automaticity properties of sequences apply to the case $(\mathbb{N}, +)$.

To simplify our language we say that $M \subset \mathbb{N}$ is p-automatic if it is H_p automatic, where $H_p(n) = pn$. A set $M \subset \mathbb{N} \times \mathbb{N}$ is called (p, q)-automatic if it is $(H_p \times H_q)$-automatic, the same applies for functions $F : \mathbb{N} \to \mathbb{N}$. We also simplify the notion of the decimations. Instead of $\partial_v^{H_p}$, we write ∂_v whenever there is no risk of confusion.

We begin with a fundamental estimate on the growth of automatic functions.

Theorem 4.4.1. *Let $G : \mathbb{N} \to \mathbb{N}$ be a (p, q)-automatic function. Then there exists a constant $K_1 > 0$ such that*

$$G(n) \leq K_1 n^{\frac{\log q}{\log p}}$$

holds for all $n \in \mathbb{N}, n \neq 0$.

Proof. Let $n = v_0 + p v_1 + \cdots + p^k v_k, v_j \in \{0, \ldots, p-1\}$ be the unique representation of n, $n \neq 0$. Then the reduced kernel graph of $\chi_{\mathrm{Gr}(G)}$ has the unique first component property, see Remark 3, p. 103. Thus there exists a unique shortest path $\mathfrak{w}_n = (\varnothing; (v_0, w_0), \ldots, (v_k, w_k), (0, w_{k+1}), \ldots (0, w_{k+l})) \in \mathfrak{W}_n^*$, see Remark 1, p. 103, where $l = l(n)$ such that $w_{k+l} \neq 0$. Since the reduced kernel graph is finite and the path is shortest there exists an upper bound L for all $l(n)$, $n \in \mathbb{N}$. Thus we conclude that

$$G(n) \leq q^{k+L+1},$$

where k depends on n. Since $w_k \neq 0$, we have the important relation

$$k \leq \frac{\log n}{\log p}$$

and we obtain

$$G(n) \leq q^{L+1} n^{\frac{\log q}{\log p}},$$

the desired result. $\square$

Remarks.
  1. In terms of groups, p- and q-valuation maps, the above theorem reads as follows. Let $G : \Gamma_1 \to \Gamma_2$ be $(H_1 \times H_2)$-automatic and let V_1 and V_2 be complete digit sets for H_1 and H_2, respectively. If the cardinality of V_1 is equal to p and the cardinality of V_2 is equal to q, then there exists a constant K such that

$$\mathfrak{V}_q(G(\gamma)) \leq K \, \mathfrak{V}_p(\gamma)^{\frac{\log q}{\log p}}$$

holds for all $\gamma \neq e_1$.

2. If one wants to take also the zero into account, the estimate becomes

$$G(n) \leq K(n+1)^{\frac{\log q}{\log p}},$$

where K is a properly chosen positive number.

3. The above estimate on the growth readily generalizes to the case where $M \subset \mathbb{N}$ and $G : M \to \mathbb{N}$ is a (p, q)-automatic map. We then have

$$G(m) \leq K_1 m^{\frac{\log q}{\log p}}$$

for all $m \in M$.

As a simple consequence we note

Corollary 4.4.2. *Let $G : \mathbb{N} \to \mathbb{N}$ be (p, q)-automatic. If $p > q$, then G is not injective.*

Proof. By our basic inequality we have that

$$G(n) \leq K(n+1)^{\frac{\log q}{\log p}} < n$$

for all $n \geq n_0$. Since G is a function from $\mathbb{N}$ to $\mathbb{N}$, it cannot be injective. $\square$

Corollary 4.4.3. *Let $G : \mathbb{N} \to \mathbb{N}$ be (p, q)-automatic. If $p < q$ and G is injective, then G is not surjective.*

Proof. Suppose that there exist an automatic function $G : \mathbb{N} \to \mathbb{N}$ which is bijective. Then by Lemma 4.2.3, G^{-1} would be (q, p) automatic and bijective which contradicts Corollary 4.4.2. $\square$

Examples.

1. The identity on $\mathbb{N}$ is a (p, p)-automatic map for all $p \geq 2$.

2. If p and q are different, then the identity on $\mathbb{N}$ is not (p, q)-automatic.

3. If $\alpha, \beta \in \mathbb{N}$, then the function $n \mapsto \alpha n + \beta$ is (p, p)-automatic for all $p \geq 2$. See Lemma 4.2.2.

4. If $p > q$, then there exists a surjective automatic map $G : \mathbb{N} \to \mathbb{N}$. Figure 4.7 shows the reduced kernel graph of G, where the residue sets are $V_p = \{0, 1, \ldots, p-1\}$ and $V_q = \{0, 1, \ldots, q-1\}$ and $\nu : V_p \to V_q$ is any surjective map with $\nu(0) = 0$. Note that G even has the transducer property and G is not injective.

$$(0,0)$$
$$(i, v(i))_{i=1}^{p-1}$$

Figure 4.7. Reduced kernel graph of a surjective (p, q)-automatic function.

5. Let $M = \{3^n \mid n \in \mathbb{N}\}$ then $M \subset \mathbb{N}$ is 3-automatic. There exists no $(2, 3)$-automatic function $G : \mathbb{N} \to \mathbb{N}$ such that $G(\mathbb{N}) = M$ and G is injective.

Let us assume that G is an injective function such that $G(\mathbb{N}) = M$. By Theorem 4.4.1, we have $G(n) \leq Kn^{\frac{\log 3}{\log 2}}$. For every $n \in \mathbb{N}$ there exists $n_0(n) \in \mathbb{N}$ such that

$$n_0(n) \leq \frac{2^n}{K^{\frac{\log 2}{\log 3}}} < n_0 + 1.$$

Then $G(n_0(n)) \leq 3^n$. Since G is injective, $G([0, n_0(0)]) \subset (M \cap [0, 3^n])$ contains $n_0(n) + 1$ elements. If n is sufficiently large, then $n_0(n) > n$ which is a contradiction.

On the other hand, if we drop the requirement of injectivity, then it is possible to construct a $(2, 3)$-automatic map with $G(\mathbb{N}) = \mathbb{N}$. The reduced kernel graph of the graph of such a function is shown in Figure 4.8. Note that the kernel graph also has the transducer property.

$$(0, 0)$$
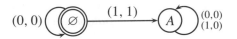
$$(1, 1)$$
$$A \quad (0,0)\ (1,0)$$

Figure 4.8. Reduced kernel graph for the $(2, 3)$-automatic set $\mathbb{N} \times \{3^n \mid n \in \mathbb{N}\}$.

The next theorem is concerned with a lower bound for (p, q)-automatic functions. Since any constant function is automatic, no useful lower bound can be achieved without certain restrictions on the automatic function.

Definition 4.4.4. A function $F : \mathbb{N} \to \mathbb{N}$ has *the finite preimage property* if the set $F^{-1}(m) = \{n \mid F(n) = m\}$ is finite for all $m \in \mathbb{N}$.

Examples 3 and 4 from above both provide examples for automatic functions with the finite preimage property.

Theorem 4.4.5. *Let $G : \mathbb{N} \to \mathbb{N}$ be a (p, q)-automatic function such that G has the finite preimage property. Then there exists an $n_0 \in \mathbb{N}$ and a constant $K_2 > 0$ such that*

$$K_2\, n^{\frac{\log q}{\log p}} \leq G(n)$$

holds for all $n \geq n_0$.

Proof. For the proof we construct the $(*, 0)$-skeleton of the (p, q)-kernel graph of $\chi_{Gr(G)}$ by removing all edges of the form $\underline{h} \xrightarrow{(v,w)} g$ with $w \neq 0$.

Now let $\underline{h} \in \ker(\chi_{Gr(G)})$ such that $\underline{h}(0, 0) = 1$ and such that there exists a $g \in \ker(\chi_{Gr(G)})$ with $\partial_{(v,0)}(\underline{g}) = \underline{h}$ and $v \neq 0$.

Then the number of paths in the $(*, 0)$-skeleton leading to g is finite. Indeed, if we suppose that the number of paths leading to g is infinite, then there exists a cycle in the $(*, 0)$-skeleton, see Figure 4.9.

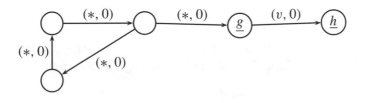

Figure 4.9. A cycle in the $(*, 0)$-skeleton.

This shows that there are infinitely many paths in the kernel graph which start at $\chi_{Gr(G)}$ and terminate at $\underline{h}$. The first component of each of these paths corresponds to different values of n while the second component remains constant. Therefore there exists an $n \in \mathbb{N}$ such that $G^{-1}(n)$ is an infinite set. This is a contradiction.

It follows that there exists an $L \in N$ such that for all $\underline{h} \in \ker(\chi_{Gr(G)})$, where $\underline{h}(0, 0) = 1$ and g with $\partial_{(v,0)}(\underline{g}) = \underline{h}$, $v \neq 0$, the length of the longest path leading to g in the $(*, 0)$-skeleton has length at most L.

Now let $(n, G(n))$, $n \neq 0$, be given and let $(v_0, w_0), \ldots, (v_k, w_k)$ be the unique representation of $(n, G(n))$. We distinguish two cases.

- If $w_k \neq 0$, then it follows that $G(n) \geq q^k$, and using the fact that

$$k \geq \frac{\log n}{\log p} - 1$$

  one obtains

$$G(n) \geq \frac{1}{q} n^{\frac{\log q}{\log p}}.$$

- If $w_k = 0$, then $v_k \neq 0$. If $k \geq L$, then, by our above considerations, at least one of the $w_{k-L}, w_{k-L+1}, \ldots, w_k$ is different from 0. This gives

$$G(n)_q \geq q^{k-L} \geq \frac{1}{q^{L+1}} n^{\frac{\log q}{\log p}},$$

  where we also used the inequality $k \geq \frac{\log n}{\log p} - 1$.

- The finitely many cases left, namely $w_k = 0$ and $k < L$ for the representation of $(n, G(n))$ and $(0, G(0))$ are incorporated in the constant n_0.

These three cases complete the proof.   □

Remarks.

1. If we assume that G satisfies the assumptions of Theorem 4.4.5 and $G^{-1}(0) = 0$, then there exists a $K > 0$ such that

$$G(n) \geq K \, n^{\frac{\log q}{\log p}}$$

   holds for all $n \in \mathbb{N}$.

2. As for Theorem 4.4.1, the above theorem also applies for (p, q)-automatic maps $G : M \to \mathbb{N}$ provided G has the finite preimage property and $M \subset \mathbb{N}$ is a p-automatic subset.

As a first application of Theorem 4.4.5 we can generalize Corollary 4.4.3.

Corollary 4.4.6. *Let $G : \mathbb{N} \to \mathbb{N}$ be a (p, q)-automatic map. If $p < q$ and G has the finite preimage property, then G is not surjective.*

Proof. By Theorem 4.4.5, there exists a constant such that

$$G(n) \geq K_2 n^{\frac{\log q}{\log p}}$$

holds for all $n \geq n_0$. Thus we see that

$$\left| G(\{0, \ldots, n\}) \cap \{0, \ldots, [K_2 n^{\frac{\log q}{\log p}}]\} \right| \leq n$$

for all $n \geq n_0$. Since $p < q$, it follows that $n < K n^{\frac{\log q}{\log p}}$ for n sufficiently large. This shows that G is not injective.   □

Remark. The assertion of the above corollary remains true if we consider (p, q)-automatic maps $G : M \to \mathbb{N}$.

As a further application of Theorems 4.4.1, 4.4.5 we consider the usual addition and multiplication on the natural numbers. In view of Theorem 4.2.6 and of Lemma 4.2.8 it is natural to ask whether the addition or multiplication are (p, q, r)-automatic. We begin with the addition.

Theorem 4.4.7. *The set* $\mathrm{Gr}(\mathrm{Add}) = \{(m, n, \mathrm{Add}(m, n)) \mid m, n \in \mathbb{N}\} \subset \mathbb{N}^3$, *i.e. the graph of the map* $\mathrm{Add}(n, m) = n + m \in \mathbb{N}$, *is (p, q, r)-automatic if and only if* $p = q = r \geq 2$.

Proof. Suppose that Gr(Add) is (p, q, r)-automatic. Since the set $V_p \times V_q$ is a complete digit set for $\mathbb{N}^2$ w.r.t. $H_q \times H_p$, we define a pq-valuation $\mathfrak{V}_{pq} : \mathbb{N}^2 \to \mathbb{N}$ as the unique continuation of

$$\mathfrak{V}_{pq}((i, j)) = i + pj$$

for $(i, j) \in V_p \times V_q$. The function $G = \text{Add} \circ \mathfrak{V}_{pq}^{-1} : \mathbb{N} \to \mathbb{N}$ is well defined and (pq, r)-automatic. Furthermore, G has the finite preimage property and $G^{-1}(0) = 0$. By Theorems 4.4.1, 4.4.5 there exist constants K_1 and K_2 such that

$$K_1 n^{\frac{\log r}{\log pq}} \leq G(n) \leq K_2 n^{\frac{\log r}{\log pq}}$$

holds for all $n \in \mathbb{N}$.

Since $\mathfrak{V}_{pq}$ is a pq-valuation one computes that

$$\mathfrak{V}_{pq}(p^n, 0) = (pq)^n \text{ and } \mathfrak{V}_{pq}(0, q^n) = p(pq)^n.$$

This gives the following values for $G = \text{Add} \circ \mathfrak{V}_{pq}^{-1}$:

$$G((pq)^n) = p^n \text{ and } G(p(pq)^n) = q^n.$$

In view of the inequalities satisfied by G, we therefore get the two inequalities

$$K_1 r^n \leq p^n \leq K_2 r^n$$
$$K_1 p^{\frac{\log r}{\log pq}} r^n \leq q^n \leq K_2 p^{\frac{\log r}{\log pq}} r^n$$

for all $n \in \mathbb{N}$. This is only possible if $p = q = r$. It remains to prove that $\chi_{\text{Gr(Add)}}$ is (p, p, p)-automatic for $p \geq 2$. This follows immediately from the invariance of the set

$$K = \left\{\underline{0}, \sum_{n,m}(n, m, n + m), \sum_{n,m \in \mathbb{N}}(n, m, n + m + 1)\right\} \subset \Sigma(\mathbb{N}^3, \mathbb{B})$$

under the decimations $\partial_{(i,j,k)}$, $(i, j, k) \in V_p \times V_q \times V_r$. $\square$

As an important consequence of the above theorem we have

Corollary 4.4.8. *If $F, G : \mathbb{N} \to \mathbb{N}$ are both (p, q)-automatic functions, then the sum $(F + G)(n) = F(n) + G(n)$ is also a (p, q)-automatic function.*

As far as the multiplication is concerned we state

Theorem 4.4.9. *There exists no $(p, q, r) \in \mathbb{N}^3$, $p, q, r \geq 2$ such that the set $\text{Gr(Mul)} = \{(n, m, nm) \mid n, m \in \mathbb{N}\}$, i.e., the graph of the multiplication, is (p, q, r)-automatic.*

Proof. The proof is analogous to the proof in case of addition. In order to avoid complications with 0 we consider the graph of the multiplication only for positive natural numbers.

We assume that there exist (p, q, r) such that Gr(Mul) is (p, q, r)-automatic. We consider the pq-valuation $\mathfrak{V}_{pq} : \mathbb{N}^2 \to \mathbb{N}$ defined in the proof of Theorem 4.4.7. Then the function $G = \text{Mul} \circ \mathfrak{V}_{pq}^{-1} : \mathbb{N} \setminus \{0\} \to \mathbb{N}$ is a (pq, r)-automatic map with finite preimage property.

As before one easily computes that $G((p+1)(pq)^n) = (pq)^n$ and $G((pq)^n+1) = p^n$. This leads to the inequalities

$$K_1' r^n \leq (pq)^n \leq K_2' r^n$$

$$K_1'' r^n \leq \quad p^n \quad \leq K_2'' r^n$$

for certain positive constants K_i', K_i'', $i = 1, 2$, and all $n \in \mathbb{N}$. This is impossible for any triple (p, q, r) with $p, q, r \geq 2$. $\square$

We have seen that in case $p < q$ there exists no surjective (p, q)-automatic function $G : \mathbb{N} \to \mathbb{N}$ with the finite preimage property. Thus the question arises whether there exists a surjective and (p, q)-automatic function $G : \mathbb{N} \to \mathbb{N}$ at all. The question is answered by the next theorem.

Theorem 4.4.10. *Let $p < q$. Then there exists no surjective (p, q)-automatic function $G : \mathbb{N} \to \mathbb{N}$.*

Proof. The idea of the proof is quite simple. We assume that there exists a surjective (p, q)-automatic function G. Then, by the above results, G does not have the finite preimage property. We therefore try to find a p-automatic subset $M \subset \mathbb{N}$ such that $G : M \to \mathbb{N}$ is surjective and has the finite preimage property. This yields the desired contradiction, see the remark on p. 116. The main difficulty is therefore the construction of the set M. M will be obtained by a careful modification of the reduced (p, q)-kernel graph of $\chi_{\text{Gr}(G)}$. As already seen in the proof of Theorem 4.4.5, it is the structure of the $(*, 0)$-skeleton of the reduced kernel graph which causes the existence of infinitely many preimages. In particular, cycles in the $(*, 0)$-skeleton 'create' infinitely many preimages.

An elementary cycle in the $(*, 0)$-skeleton is a path whose initial point coincides with its terminal point and for every vertex $\underline{h}$ of the path there exists a unique edge $\underline{g} \xrightarrow{(u,0)} \underline{h}$ of the path. In other words, an elementary cycle is as short as possible. Due to the finiteness of the $(*, 0)$-skeleton, there exist only finitely many elementary paths. These are denoted by $C_1, \ldots, C_L$. The length of the paths, i.e., the number of edges in the cycle, is denoted by $d_1, \ldots, d_L$.

As a next step, we construct a special (p, q)-automaton $(\mathcal{B}, \varnothing, \mathbb{B}, \omega, \{\alpha_{(a,b)} \mid a \in \{0, \ldots, p-1\}, b \in \{0, \ldots, q-1\}\})$. The set of states is given by

$$\mathcal{B} = \left(\ker_{p,q}(\chi_{\mathrm{Gr}(G)}) \times \prod_{j=1}^{L} \{0, 1, \ldots, d_L - 1\} \right) \cup \{\underline{h}^* \mid \underline{h} \in \ker_{p,q}(\chi_{\mathrm{Gr}(G)})\}.$$

The elements of $\mathcal{B}$ are denoted by $(\underline{h}, \boldsymbol{w})$ with $\boldsymbol{w} = (w_j)_{j=1}^{L}$ or as $\underline{h}^*$. The output function $\omega : \mathcal{B} \to \mathbb{B}$ is defined by

$$\omega(s) = \begin{cases} 0 & \text{if } s = \underline{h}^* \\ \underline{h}(0,0) & \text{if } s = (\underline{h}, \boldsymbol{w}). \end{cases}$$

Finally we define the transition function $\alpha_{(a,b)}$.

1. If $(a, b) \in \{0, \ldots, p-1\} \times \{1, \ldots, q-1\}$, then

$$\alpha_{(a,b)}((\underline{h}, \boldsymbol{w})) = (\partial_{(a,b)}(\underline{h}), (0, \ldots, 0))$$
$$\alpha_{(a,b)}(\underline{h}^*) = (\partial_{(a,b)}(\underline{h}), (0, \ldots, 0)).$$

2. If $(a, 0) \in \{0, \ldots, p-1\} \times \{0\}$ and $\underline{h}^* \in \mathcal{B}$, then

$$\alpha_{(a,0)}(\underline{h}^*) = \left(\partial_{(a,0)}(\underline{h}) \right)^*.$$

3. If $(a, 0) \in \{0, \ldots, p-1\} \times \{0\}$ and $(\underline{h}, \boldsymbol{w}) \in \mathcal{B}$, then we define an element $\boldsymbol{w}' = (w_j')$ in $\mathbb{N}^L$, where w_j' is given by

$$w_j' = w_j + 1,$$

if $\underline{h}$ and $\partial_{(a,0)}(\underline{h})$ are both vertices of the elementary cycle C_j of the $(*, 0)$-skeleton.

If $\partial_{(a,0)}(\underline{h})$ is not a vertex of the elementary cycle C_j, then $w_j' = 0$. If $\partial_{(a,0)}(\underline{h})$ belongs to the vertices of C_j and $\underline{h}$ does not belong to the vertices of C_j, then $w_j' = 0$. In all other cases $w_j' = w_j$. Finally, we set

$$\alpha_{(a,0)}((\underline{h}, \boldsymbol{w})) = \begin{cases} \left(\partial_{(a,0)}(\underline{h}) \right)^* & \text{if there exists } j \text{ such that } w_j' = d_j \\ (\partial_{(a,0)}(\underline{h}), \boldsymbol{w}) & \text{otherwise.} \end{cases}$$

It remains to define the distinguished element $\varnothing$. We set $\varnothing = (\chi_{\mathrm{Gr}(G)}, (0, \ldots, 0))$.

Figure 4.10 shows how a cycle of length 2 in the $(*, 0)$-skeleton of the kernel graph reappears in the extended transition graph defined by the above automaton. Note that the cycle has become a cycle in the $*$-states.

Let us pause for a moment and reflect upon the properties of the transition graph defined by the above automaton. A path starting in $\chi_{\mathrm{Gr}(G)}$ in the kernel graph of

$\mathrm{Gr}(G)$ corresponds to a path starting in $\varnothing$ in the transition graph of the above defined automaton. The vector $\boldsymbol{w}$ counts the number of successive steps done in the elementary cycle C_j, $j = 0, \ldots, L$. If the path leaves a cycle C_j, then w_j is reset to 0. If the path has completed an elementary cycle, then the path in the transition graph of the automaton enters a $*$-state $\underline{h}^*$. The only way to leave a $*$-state is via $\alpha_{(a,b)}$ with $h \neq 0$.

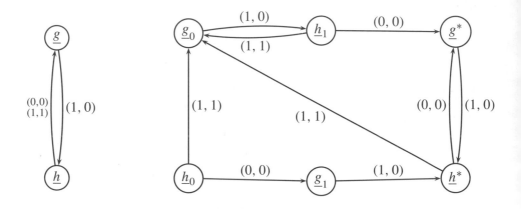

Figure 4.10. Transformation of a cycle in the $(*, 0)$-skeleton.

It is clear that the above defined automaton is a (p, q)-automaton. Since its values are in $\mathbb{B}$ it defines a subset M of $\mathbb{N} \times \mathbb{N}$. It remains to show the following three assertions.

1. The projection on the second coordinate $p_2(M)$ is equal to $\mathbb{N}$.

2. The projection on the first coordinate $p_1(M)$ is unbounded.

3. The restriction of G on $p_1(M)$ defines a function which has the finite preimage property.

Let $m \in \mathbb{N}$, $m \neq 0$ be given. Then there exists a minimal $n \in \mathbb{N}$ such that $G(n) = m$. Let $(n, m) = (\chi_{\mathrm{Gr}(G)}, ((n_0, m_0), \ldots, (n_k, m_k)))$ be the path defined by the unique representation of (n, m). Then the path terminates at a kernel element $\underline{h}$ of the reduced kernel graph of $\chi_{\mathrm{Gr}(G)}$ such that $\underline{h}(0, 0) = 1$. Now we consider the same path in the transition graph of the above defined automaton. To this end we consider the sequence of states $(s_j)_{j=0}^{k}$ defined by

$$s_j = \alpha_{(m_j, n_j)} \circ \cdots \circ \alpha_{(n_0, m_0)}(\chi_{\mathrm{Gr}(G)}).$$

If $m_k \neq 0$, then $s_k = (\underline{h}, (0, \ldots, 0))$ and $\omega(s_m) = 1$, i.e., the point (n, m) belongs to M.

If $m_k = 0$, then n_k is different from 0, for otherwise $(n_0, m_0), \ldots, (n_k, m_k)$ would not be the unique representation of (n, m). Since $m \neq 0$ there exists an $l < k$ such that $m_l \neq 0$ which means that the state s_l is not a $*$-state. It remains to show that the states s_j, $j = l + 1, \ldots, k$ are no $*$-states either. Due to the minimality of n, the path starting in s_l and following the edges $(n_{l+1}, m_{l+1}), \ldots, (n_k, m_k)$ does not contain an elementary cycle. This shows that the final state s_k is not a $*$-state and therefore $\omega(s_k) = \underline{h}(0, 0) = 1$, i.e., $(m, n) \in M$. This proves the first assertion.

A similar argument applies for the remaining case $m = 0$.

The second assertion follows directly from our above considerations. We have shown that $p_1(M)$ contains all minimal preimages of all $m \in \mathbb{N}$. This means that M is unbounded.

The last point is to show that $G : M \to \mathbb{N}$ has the finite preimage property. By our construction of the automaton, a $(*, 0)$-cycle in the transition graph of the automaton occurs only for $*$-states and $\omega(\underline{h}^*) = 0$. This yields that $G^{-1}(m) \cap M$ is finite for all $m \in \mathbb{N}$.

Therefore we have constructed a (p, q)-automatic function $G : M \to \mathbb{N}$ which is surjective and has the finite preimage property. This is impossible and therefore no surjective (p, q)-automatic function $G : \mathbb{N} \to G$ does exist. $\square$

An immediate consequence of the proof the above theorem is the following corollary.

Corollary 4.4.11. *If $G : \mathbb{N} \to \mathbb{N}$ is a (p, q)-automatic function, then there exists a p-automatic subset M of $\mathbb{N}$ such that $G(M) = G(\mathbb{N})$ and $G : M \to \mathbb{N}$ has the finite preimage property.*

If $G : \mathbb{N} \to \mathbb{N}$ is a (p, q)-automatic function with $p < q$, then, as we have seen, $G(\mathbb{N})$ is a q-automatic set different from $\mathbb{N}$. It is therefore natural to ask which q-automatic subsets M_q are of the form $M_q = G(\mathbb{N})$. In view of Corollary 4.4.11 and the lower bound for the growth of G restricted on M it is not surprising that the "density" of M_q plays a crucial role in whether M_q is of the form $M_q = G(\mathbb{N})$ or not.

Before we begin with a closer study of the "density" we demonstrate that under certain circumstances it is possible to construct a (p, q)-automatic map $G : \mathbb{N} \to M_q$ such that $G(\mathbb{N}) = M_q$.

Definition 4.4.12. Let χ be a p-automatic sequence, and let $\mathcal{G}_q^*(\chi)$ be the reduced q-kernel graph of χ. For $\underline{h} \in \ker_p^*(\chi)$ the number

$$\left| \{ j \mid \partial_j(\underline{h}) \in \ker_p^*(\chi), \ j = 0, \ldots, q - 1 \} \right|$$

is called the *degree of $\underline{h}$* and denoted by $\deg(\underline{h})$. The number

$$\max \{ \deg(\underline{h}) \mid \underline{h} \in \ker_p^*(\chi) \}$$

is called the *degree of $\mathcal{G}_p^*(\chi)$* and is denoted by $\deg(\mathcal{G}_p^*(\chi))$.

Lemma 4.4.13. *Let* $M \subset \mathbb{N}$ *be a* q-*automatic subset. If the degree of the reduced* q-*kernel graph of* $\ker_q^*(\chi_M)$ *satisfies*

$$\deg(\mathcal{G}_q^*(\chi_M)) \leq p,$$

then there exists a (p, q)-*automatic function* $G : \mathbb{N} \to M$ *such that* $G(\mathbb{N}) = M$.

Proof. We construct a (p, q)-automatic function with the transducer property. To this end, we define for every $\underline{h} \in \ker_q^*(\chi_M)$ surjective maps $v_{\underline{h}} : \{0, \ldots, p-1\} \to W_{\underline{h}}$, where $W_{\underline{h}} = \{j \mid \partial_j(\underline{h}) \in \ker_p^*(\chi_M), \ j = 0, \ldots, q-1\}$.

If $\underline{h} \in \ker_q^*(\chi_M)$ and if $\underline{h}(0) = 1$, then $v_{\underline{h}} : V \to W_{\underline{h}}$ can be any surjective map which satisfies $v_{\underline{h}}(0) = 0$.

If $\underline{h} \in \ker_q^*(\chi_M)$ and $\underline{h}(0) = 0$, then there exists a minimal $n \geq 1$ such that $\underline{h}(n) = 1$. If $n = n_0 + qn'$, then $v_{\underline{h}} : V \to W_{\underline{h}}$ is any surjective map such that $v_{\underline{h}}(0) = n_0$.

Using the maps $v_{\underline{h}}$ we define a (p, q)-automaton in the obvious way. The states are given by the q-kernel of χ_M, the basepoint is χ_M, the output function is given by $\omega(\underline{h}) = \underline{h}(0)$, and the transition functions are defined by

$$\alpha_{(v,w)}(\underline{h}) = \begin{cases} \partial_w(\underline{h}) & \text{if } v_{\underline{h}}(v) = w \\ \underline{0} & \text{otherwise,} \end{cases}$$

for $\underline{h} \in \ker_p^*(\chi_M)$, and $\alpha_{(v,w)}(\underline{0}) = \underline{0}$ for all $(v, w) \in \{0, \ldots, p-1\} \times \{0, \ldots, q-1\}$.

It remains to prove that the above automaton defines a function $G : \mathbb{N} \to M$. By Lemma 4.3.4, it suffices to show that the automaton has the unique first component property.

If $n = 0$ and $\chi_M(0) = 1$, then we have $G(0) = 0$ is well defined since $v_{\chi_M}(0) = 0$. If $\chi_M(0) \neq 0$, then there exists a smallest $m = m_0 + m_1 p + \cdots + m_k p^k, m_k \neq 0$, such that $\chi_M(m) = 1$ and we set $G(0) = m$. Due to the construction of the maps $v_{\underline{h}}$, we have that the path $(0, m_0), (0, m_1), \ldots, (0, m_k)$ is the unique path which terminates at a kernel element $\underline{h}$ with $\underline{h}(0) = 1$.

If $n = n_0 + n_1 p + \cdots + n_k p^k, n_k \neq 0$, then the maps $v_{\underline{h}}$ define a unique path $(n_0, m_0), \ldots, (n_k, m_k)$ which terminates in $\underline{h} \in \ker_q^*(\chi_M)$. If $\underline{h}(0) = 1$, then we define $G(n) = m_0 + m_1 q + \cdots + m_k p^k$. If $\underline{h}(0) = 0$, then with the same arguments as for the case $n = 0$, there exists a unique extension

$$(n_0, m_0), \ldots, (n_k, m_k), (0, m_{k+1}), \ldots (0, m_{k+l})$$

such that $m_{k+l} \neq 0$ and the path terminates in a kernel element $\underline{h}'$ with $\underline{h}'(0) = 1$. $\square$

As a corollary we note

Corollary 4.4.14. *Let* $M \subset \mathbb{N}$ *be a* q-*automatic subset. If*

$$\deg(\underline{h}) = p$$

for all $\underline{h} \in \ker_q^*(\chi_M)$ and $p \geq 2$, then there exists a bijective (p, q)-automatic function
$G : \mathbb{N} \to M$.

Examples.

1. The Thue–Morse sequence $\underline{t} = (t_n)_{n \in \mathbb{N}}$ defines a subset $M_{\underline{t}} \subset \mathbb{N}$. In Figure 3.1 the reduced kernel graph of $\underline{t}$ is shown. Since every vertex in the reduced kernel graph has degree 2, it follows that there exists a $(2, 2)$-automatic function $G : \mathbb{N} \to M_{\underline{t}}$. Following the procedure in the proof of Lemma 4.4.13 we construct the reduced kernel graph of $\chi = \chi_{\mathrm{Gr}(G)}$. It is shown in Figure 4.11. See also Example 3, p. 108, for further properties of G.

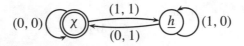

Figure 4.11. Transition graph for the bijective function $G : \mathbb{N} \to M$, where M is the support of the Thue–Morse sequence.

2. Figure 4.12 shows the reduced kernel graph of a $(2, 3)$-automatic function $G : \mathbb{N} \to M$, where M is the set of integers n which have no 1 in their 3-adic representation.

Figure 4.12. Transition graph for the bijective function $G : \mathbb{N} \to M$. M is the set of integers without a 1 in the 3-adic representation.

We now start our investigation on the "density" of the set $M_q \subset \mathbb{N}$. To this end we introduce certain counting functions.

Definition 4.4.15. Let $M \subset \mathbb{N}$ be a p-automatic set. The function $c(M) : \mathbb{N} \to \mathbb{N}$ defined by

$$c(M)(n) = |[0, n+1[\cap M|$$

is called the *counting function* of M. The function $C_p(M) : \mathbb{N} \to \mathbb{N}$ defined by

$$C_p(M)(n) = |[0, p^n[\cap M|$$

is called the *p-counting function*.

Examples.

1. $C_p(\mathbb{N})(n) = p^n$ and $c(\mathbb{N})(n) = n + 1$.

2. Consider the Thue–Morse sequence (t_n) as a characteristic sequence of a subset $M_t \in \mathbb{N}$. The 2-counting function $C_2(M)$ is given by

$$C_2(M_t)(n) = 2^{n-1}$$

for all $n \geq 1$.

3. If M is a bounded set, then

$$c(M)(n) = |M|$$

for all $n \geq n_0$.

4. If $M = \{2^n \mid n \in \mathbb{N}\}$, then
$$C_2(M)(n) = n$$

for all n.

The following lemma is immediate. It provides a first hint how the density of M_q influences the existence of a surjective function $G : \mathbb{N} \to M_q$.

Lemma 4.4.16. *Let $M_q \subset \mathbb{N}$ be a q-automatic subset and let $G : \mathbb{N} \to M_q$ be a (p, q)-automatic map with the finite preimage property. Let K_2 be the constant given in Theorem 4.4.5. If there exists an n_0 such that*

$$C_q(M_q)(n) > \left(\frac{1}{K_2}\right)^{\frac{\log p}{\log q}} p^n$$

holds for all $n \geq n_0$, then $G : \mathbb{N} \to M_q$ is not surjective.

Proof. Assume that there exist a (p, q)-automatic function $G : \mathbb{N} \to M_q$ such that G is surjective and has the finite preimage property.

If $n \geq K_2^{-\frac{\log p}{\log q}} p^m$ and $m \geq m_0$ for a certain large m_0, then it follows that

$$G(n) \geq K_2 n^{\frac{\log q}{\log p}} \geq q^m.$$

This shows that $C_q(M_q)(m) \leq K_2^{-\frac{\log p}{\log q}} p^m$ for all m which proves the assertion.   □

The counting function $c(M)$ of a q-automatic subset M allows us to establish the existence of a very special (p, q)-automatic function $G : \mathbb{N} \to M$ such that $G(\mathbb{N}) = M$. We have the following surprising result.

Theorem 4.4.17. *Let $M \subset \mathbb{N}$ be an unbounded q-automatic subset. Then $c(M) : \mathbb{N} \to \mathbb{N}$ is (q, p)-automatic if and only if there exists a (p, q)-automatic function $F : \mathbb{N} \to M$ such that F is bijective and monotone increasing.*

Proof. Let $F : \mathbb{N} \to M$ be (p, q)-automatic with the above properties. By purely set theoretic operations we construct the graph of the counting function $c(M)$ from the graph of F.

For a subset $L \subset \mathbb{N} \times \mathbb{N}$, we define the upper closure L^{up} of L by

$$L^{\text{up}} = \bigcup_{(l_1, l_2) \in L} \{(l_1, n) \mid n \geq l_2\}.$$

The upper closure $\text{Gr}(F)^{\text{up}}$ of $\text{Gr}(F)$ is then given by

$$\text{Gr}(F)^{\text{up}} = \{(n, l) \mid n \in \mathbb{N} \text{ and } l \geq F(n)\}.$$

Then $\text{Gr}(F)^{\text{up}}$ is a (p, q)-automatic subset. It is easy to see that

$$\chi_{\text{Gr}(F)^{\text{up}}} = \chi_{\text{Gr}(F)} \cdot \chi_{\{0\} \times \mathbb{N}} = \bigvee_{n \in \mathbb{N}} x^n y^{F(n)} \cdot \bigvee_{n \in \mathbb{N}} y^n,$$

i.e., $\chi_{\text{Gr}(F)^{\text{up}}}$ is the Cauchy product of two elements in $\Sigma(\mathbb{N}^2, \mathbb{B})$. Since both factors are (p, q)-automatic, it follows from Theorem 4.1.7 that $\text{Gr}(F)^{\text{up}}$ is (p, q)-automatic.

Now let $G : \mathbb{N} \to M$ be defined by $G(n) = F(n + 1)$. Then G is also (p, q)-automatic. This follows from the fact that the addition is p-automatic and that compositions of automatic functions are automatic. Then the sequence

$$\chi = \chi_{\text{Gr}(F)^{\text{up}} \triangle \text{Gr}(G)^{\text{up}}}$$

which is the characteristic sequence of the symmetric difference of $\text{Gr}(F)^{\text{up}}$ and $\text{Gr}(G)^{\text{up}}$ is (p, q)-automatic, see 4. of Lemma 4.1.3. Since F is monotone increasing, it follows that

$$\text{supp}(\chi)^T = \{(n, l) \mid \chi(l, n) = 1\} = \text{Gr}(c(M)).$$

This proves the (q, p)-automaticity of the counting function of M.

Let us now assume that $c(M) : \mathbb{N} \to \mathbb{N}$ is (q, p)-automatic. Since M is unbounded, it follows that $c(M)$ is surjective. Moreover, for $m \in M$ the set

$$(\mathbb{N} \times \{m\}) \cap \text{Gr}(c(M))^T$$

contains only one point and, furthermore, for any $n \in \mathbb{N}$ there exists a unique $m_n \in M$ such that

$$(\mathbb{N} \times \{m_n\}) \cap \text{Gr}(c(M))^T = (n, m_n).$$

Thus the map $F : \mathbb{N} \to M$ defined by $F(n) = m_n$ provides a monotone bijection. By the construction of F we have

$$\text{Gr}(F) = (\mathbb{N} \times M) \cap \text{Gr}(c(M))^T.$$

It follows that F is (p, q)-automatic. $\qquad\square$

Examples.

1. The set $M = \{3n + 1 \mid n \in \mathbb{N}\}$ is a 2-automatic subset of $\mathbb{N}$ and its counting function is $(2, 2)$-automatic since $F(n) = 3n + 1$ provides an automatic and monotone increasing function from $\mathbb{N}$ to M.

2. Let $M = \{2^n \mid n \in \mathbb{N}\}$, then M is a 2-automatic subset of $\mathbb{N}$ and $c(M)$ is not $(2, 2)$-automatic. This follows almost immediately from the fact that $c_1 \log_2(n) \le c(M)(n) \le c_2 \log_2(n)$ holds for all $n \ge 1$ and constants $c_1, c_2 > 0$ and the growth theorem for automatic functions. Thus there exists no monotone increasing surjective and $(2, 2)$-automatic function $F : \mathbb{N} \to M$. In particular, $f(n) = 2^n$ is not a $(2, 2)$-automatic function. However, in Figure 4.13 the reduced kernel graph of a surjective $(2, 2)$-automatic function $F : \mathbb{N} \to M$ is shown. Observe that $\omega(A) = 0$, $\omega(B) = 1$.

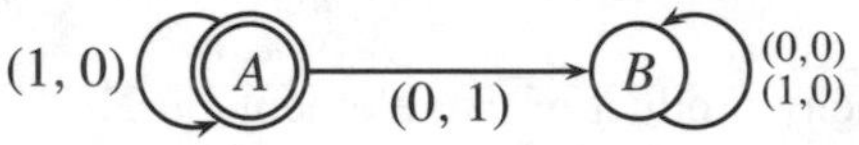

$$(1, 0) \; \overset{\curvearrowleft}{\boxed{A}} \; \xrightarrow{\;(0, 1)\;} \; \boxed{B} \; \overset{\curvearrowright}{} \; \begin{matrix}(0, 0)\\(1, 0)\end{matrix}$$

Figure 4.13. A surjective function $F : \mathbb{N} \to \{2^n \mid n \in \mathbb{N}\}$.

Note that F does not have the finite preimage property. Indeed, $F(2n + 1) = 1$ for all $n \in \mathbb{N}$.

3. The Thue–Morse sequence $\underline{t} = (t_n)_{n \ge 0} \in \Sigma(\mathbb{N}, \mathbb{B})$ provides a non-trivial example of a 2-automatic subset $M_{\underline{t}} = \{n \mid t_n = 1\}$ with a $(2, 2)$-automatic counting function $c(M)$. In fact, it follows from the construction of the Thue–Morse sequence by a substitution that

$$c(M)(n) = \begin{cases} k + 1 & \text{if } n = 2k + 1 \\ k + t_k & \text{if } n = 2k \end{cases}$$

holds for all $n \in \mathbb{N}$. From these formulas it follows that $c(M)$ is $(2, 2)$-automatic. Moreover, the monotone bijective function $F : \mathbb{N} \to M_{\underline{t}}$ is given by

$$F(n) = 2n + 1 - t_n$$

for $n \in \mathbb{N}$. From these formulas it follows that $c(M)$ is $(2, 2)$-automatic.

In order to compute the counting functions $c(M)$ and $C_p(M)$ of a p-automatic subset $M \ne \emptyset$ we use the fact that the sequence $\chi_M \in \Sigma(\mathbb{N}, \mathbb{B})$ is generated by a p-substitution $S : \Sigma(\mathbb{N}, \mathbb{B}^N) \to \Sigma(\mathbb{N}, \mathbb{B}^N)$, where N denotes the cardinality of $\ker_p(\chi_M)$. If we enumerate the kernel elements such that $\chi_M = \underline{f}_1$, then the substitution is given by its substitution polynomial

$$P_S = \sum_{j=0}^{p-1} x^j A_j,$$

where A_j are $N \times N$-matrices with entries in $\{0, 1\}$, see Definition 2.2.28. Moreover we have that the sequence $\underline{F}_M \in \Sigma(\mathbb{N}, \mathbb{B}^N)$ defined by

$$\underline{F}_M(n) = (\underline{f}_1(n), \ldots \underline{f}_N(n))$$

is a fixed point of the substitution S, i.e.,

$$\underline{F}_M = P_S H_*(\underline{F}),$$

where $H(x^j) = x^{pj}$. Thus we can write $\underline{F}_M(x) = P_S(x)\underline{F}_M(x^p)$, considered as a Cauchy product over $\mathbb{B}$. Actually, we can consider this product as the usual product over $\mathbb{Z}$ which we do from now on.

Definition 4.4.18. Let $M \subset \mathbb{N}$ be a p-automatic subset and let $\underline{f}_1, \ldots, \underline{f}_{N+1}$ be a enumeration of the kernel elements such that $\underline{f}_1 = \chi_M$ and $\underline{f}_{N+1} = \underline{0}$. Then the polynomial

$$P_S^{\text{red}} = \sum_{j=0}^{p-1} x^j A_j^{\text{red}},$$

where A_j^{red} is an $N \times N$-matrix obtained from A_j by removing the $N+1$-st row and the $N+1$-st column, is called the *reduced substitution polynomial* of M.

In case that $\underline{0} \notin \ker_p(\chi_M)$ we consider the substitution polynomial also as the reduced substitution polynomial. The next lemma justifies the above definition.

Lemma 4.4.19. *Let $M \subset \mathbb{N}$ be p-automatic and let P_S^{red} be the reduced substitution polynomial of M. Furthermore, let $\underline{F}_M^{\text{red}} \in \Sigma(\mathbb{N}, \mathbb{B}^N)$ be defined by*

$$\underline{F}_M^{\text{red}}(n) = (\underline{f}_1(n), \ldots \underline{f}_N(n)).$$

Then $\underline{F}_M^{\text{red}}$ satisfies the equation

$$\underline{F}_M^{\text{red}} = P_S^{\text{red}} H_*(\underline{F}_M^{\text{red}})$$

in $\mathbb{B}^N$ and $H : \mathbb{N} \to \mathbb{N}$ is given by $H(x^j) = x^{pj}$.

Proof. There is nothing to show if $\underline{0} \notin \ker_p(\chi_M)$. Therefore let us assume that $\underline{0}$ is a kernel element. Then the matrices A_j, $j = 0, \ldots, p-1$ have the following structure:

$$A_j = \begin{pmatrix} A_j^{\text{red}} & b \\ 0 \ldots 0 & 1 \end{pmatrix},$$

where $b \in \mathbb{B}^N$, and each row of A_j contains one 1. We now notice that $\underline{F}_M(n)$ is an element of $\mathbb{B}^N \times \{0\}$ and $A_j(a_1, \ldots, a_N, 0)^T = (b_1, \ldots, b_N, 0)^T$ holds for all

$a_1, \ldots, a_N \in \{0, 1\}$. Then the assertion follows from

$$\begin{pmatrix} A_j^{\text{red}} & b \\ 0 \ldots 0 & 1 \end{pmatrix} \begin{pmatrix} a_1 \\ \vdots \\ a_N \\ 0 \end{pmatrix} = \left(A_j^{\text{red}} \begin{pmatrix} a_1 \\ \vdots \\ a_N \\ 0 \end{pmatrix} \right),$$

where the matrix multiplication on the r.h.s. of the equation is an operation in $\mathbb{B}$.   □

The above lemma remains true if $\mathbb{B}$ is replaced by $\mathbb{Z}$.

Examples.

1. Let $(t_n)_{n \geq 0} \in \Sigma(\mathbb{N}, \mathbb{B})$ be the Thue–Morse sequence. Then the reduced substitution polynomial is given by

$$P_S = \begin{pmatrix} 1 & 0 \\ 0 & 1 \end{pmatrix} + x \begin{pmatrix} 0 & 1 \\ 1 & 0 \end{pmatrix},$$

which is equal to the substitution polynomial.

2. Let χ be the characteristic sequence of the set $\{2^n \mid n \in \mathbb{N}\}$. The sequence χ is two automatic and its reduced substitution polynomial is given by

$$P_S^{\text{red}} = \begin{pmatrix} 1 & 0 \\ 0 & 1 \end{pmatrix} + x \begin{pmatrix} 0 & 1 \\ 0 & 0 \end{pmatrix}.$$

3. For the paperfolding sequence the reduced substitution polynomial is given by

$$P_S^{\text{red}} = \begin{pmatrix} 0 & 1 & 0 \\ 0 & 0 & 1 \\ 0 & 0 & 1 \end{pmatrix} + x \begin{pmatrix} 1 & 0 & 0 \\ 0 & 0 & 1 \\ 0 & 0 & 1 \end{pmatrix}.$$

Remarks.

1. If $P(x)$ denotes the reduced p-substitution polynomial of a sequence χ_M, then $P(x)P(x^p)$ defines a p^2-substitution which has $\underline{F}_M$ as a fixed point, i.e., we have

$$\underline{F}_M(x) = P(x)P(x^p)\underline{F}_M(x^{p^2}).$$

2. The polynomial $P(x)P(x^p)$ is not necessarily equal to the reduced p^2-substitution polynomial.

We are now prepared to compute the counting function $c(M)$ and the p-counting function $C_p(M)$. The following observation is crucial for the computation. Let M be a subset of $\mathbb{N}$ and let χ_M be its characteristic sequence, i.e.,

$$\chi_M = \bigvee_{j-0}^{\infty} x^j \, \chi_M(j).$$

If we consider the formal series for χ_M as an element of the ring $\mathbb{Z}[[x]]$ of formal power series with coefficients in the ring $\mathbb{Z}$, then we have the identity

$$c(M)(x) = \sum_{j=0}^{\infty} x^j \, c(M)(j) = \frac{1}{1-x} \sum_{j=0}^{\infty} x^j \, \chi_M(j) = \frac{1}{1-x} \chi_M(x).$$

In other words, the generating function $c(M)(x)$ of the sequence $(c(M)(n))_{n \geq 0}$ is given by the above identity.

This observation yields the following lemma.

Lemma 4.4.20. *Let $M \subset \mathbb{N}$ be a p-automatic subset and let P_S^{red} denote the re-duced substitution polynomial with respect to an enumeration of the kernel elements, $\underline{f}_1 = \chi_M, \dots, \underline{f}_N$. Then the generating function $C_M(x) \in \Sigma(\mathbb{N}, \mathbb{Z}^N)$ of the sequence $((c(\mathrm{supp}(\underline{f}_1)), \dots, c(\mathrm{supp}(\underline{f}_N)))_{n \geq 0}$ with values in $\mathbb{N}^N$ satisfies the following equa-tion*

$$C_M(x) = (1 + x + \cdots + x^{p-1}) P_S^{\mathrm{red}}(x) C_M(x^p)$$

over $\mathbb{Z}$.

Proof. Since M is automatic it follows from Lemma 4.4.19 that

$$\underline{F}_M^{\mathrm{red}} = P_S^{\mathrm{red}} H_*(\underline{F}_M^{\mathrm{red}}).$$

If we consider the above equation as an equation in the ring $\mathbb{Z}$, then the equation remains true for $\underline{F}_M$. This follows from the fact that due to the construction of the matrices A_j^{red} contain at most one 1 in each row.

Thus $\underline{F}_M(x)$ satisfies

$$\underline{F}_M(x) = P_S^{\mathrm{red}}(x) \underline{F}_M(x^p)$$

viewed as an equation with coefficients in $\mathbb{Z}$. This gives

$$\frac{1}{1-x} \underline{F}_M(x) = \frac{1}{1-x} P_S^{\mathrm{red}}(x) \underline{F}_M(x^p),$$

where the multiplication by $\frac{1}{1-x}$ is defined component-wise. Hence

$$C_M(x) = \frac{1}{1-x} \underline{F}_M(x) = (1 + x + \cdots + x^{p-1}) P_S^{\mathrm{red}}(x) C_M(x^p). \qquad \square$$

Examples.

1. For the Thue–Morse sequence $\underline{t}$ we obtain the equation

$$C(x) = (1 + x)\begin{pmatrix} 1 & x \\ x & 1 \end{pmatrix}C(x^2)$$

for the counting functions of the kernel elements. If one further observes that $c(\mathrm{supp}(\underline{t}))(n) + c(\mathrm{supp}(\partial_1(\underline{t})))(n) = n$ holds for all $n \in \mathbb{N}$ we get the following equation (over $\mathbb{Z}$!) for the generating function $c(M)(x)$ with $M = \mathrm{supp}(\underline{t})$:

$$c(M)(x) = (1 - x^2)c(M)(x^2) + \frac{x}{1 - x - x^2 + x^3}.$$

The above equation yields the formulas given for $c(M)(n)$ in Example 3, p. 126.

2. If we consider the counting function for the paper folding sequence, then we obtain the equation

$$c(M)(x) = x(1 + x)c(x^2) + \frac{1}{1 - x - x^4 + x^5}$$

for the generating function of the counting function of $M = \mathrm{supp}(\underline{pf})$. The above functional equation yields the following recursive relations:

$$c(M)(2n) = c(M)(n - 1) + \left[\tfrac{2n+4}{4}\right]$$

$$c(M)(2n + 1) = c(M)(n) + \left[\tfrac{2n+5}{4}\right].$$

It is not known whether $c(M)$ is a $(2, 2)$-automatic function.

The reduced substitution polynomial also provides a tool to compute the p-counting function $C_p(M)$ for an automatic subset M.

Lemma 4.4.21. *Let M be a p-automatic subset and P_S^{red} the reduced substitution polynomial w.r.t. an enumeration of the kernel elements of χ_M. Then we have*

$$C_p(M)(n) = (1, 0, \ldots, 0)(P_S^{\mathrm{red}}(1))^n \begin{pmatrix} \underline{f}_1(0) \\ \vdots \\ \underline{f}_N(0) \end{pmatrix}$$

for all $n \in \mathbb{N}$, where we interpret $P_S^{\mathrm{red}}(1)$ as an $(N \times N)$-matrix with coefficients in $\mathbb{Z}$ after substituting 1 for x.

Proof. The proof follows from the fact that the sequence $\underline{F}_M$ is obtained as a limit of the substitution S, i.e.,

$$\underline{F} = \lim_{n \to \infty} S^n \begin{pmatrix} \underline{f}_1(0) \\ \vdots \\ \underline{f}_N(0) \end{pmatrix},$$

where the limit is in $\Sigma(\mathbb{N}, \mathbb{B}^N)$. We thus obtain in terms of formal power series over $\mathbb{Z}$

$$S^n \begin{pmatrix} \underline{f}_1(0) \\ \vdots \\ \underline{f}_N(0) \end{pmatrix} (x) = P_S^{\text{red}}(x) P_S^{\text{red}}(x^p) \ldots P_S^{\text{red}}(x^{p^n}) \begin{pmatrix} \underline{f}_1(0) \\ \vdots \\ \underline{f}_N(0) \end{pmatrix}$$

and by substituting $x = 1$ in the r.h.s. of the equation we achieve the desired result. $\square$

We have almost reached our goal, namely to characterize q-automatic subsets M which are an image of a (p, q)-automatic map $F : \mathbb{N} \to \mathbb{N}$.

There are basically two approaches to decide whether a q-automatic set is the image of a (p, q)-automatic map. The first approach uses Lemma 4.4.16 and Lemma 4.4.21. The second approach uses Lemma 4.4.13.

Examples.

1. In Figure 4.14 a reduced 3-kernel graph of a 3-automatic subset M is shown. Note that $\underline{f}(0) = \underline{g}(0) = 1$.

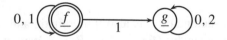

Figure 4.14. The automaton generates a set $M \subset \mathbb{N}$ which is not of the form $F(\mathbb{N})$ for a $(2, 3)$-automatic function F.

The reduced kernel polynomial is given by

$$P(x) = \begin{pmatrix} 0 & 1 \\ 0 & 1 \end{pmatrix} + x \begin{pmatrix} 1 & 0 \\ 0 & 0 \end{pmatrix} + x^2 \begin{pmatrix} 0 & 1 \\ 0 & 1 \end{pmatrix},$$

which gives

$$P(1) = \begin{pmatrix} 2 & 1 \\ 0 & 2 \end{pmatrix}.$$

Using the above result we easily compute that $C_3(M)(n) = 2^n + n2^{n-1}$. By Lemma 4.4.16, it follows that there exists no $(2, 3)$-automatic function F such that $F(\mathbb{N}) = M$ and F has the finite preimage property. Indeed, if we suppose that there exists a $(2, 3)$-automatic F with $F(\mathbb{N}) = M$ and F does not have the finite preimage property, then by Lemma 4.4.11 there exists a 2-automatic subset M_2 such that $F : M_2 \to M$ is surjective and has the finite preimage property which is a contradiction to Lemma 4.4.16.

By Lemma 4.4.13, for all $p \geq 3$ there exists a $(p, 3)$-automatic map F with $F(\mathbb{N}) = M$.

2. Figure 4.15 shows the reduced 3-kernel graph of a 3-automatic subset $M = \mathrm{supp}(f)$, where $\underline{f}(0) = \underline{g}(0) = 0$ and $\underline{h}(0) = 1$. We show that M is an image of a $(\overline{2}, 3)$-automatic map F.

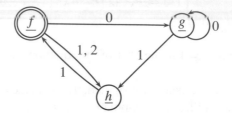

Figure 4.15. 3-automatic subset $M \subset \mathbb{N}$ which is of the form $M = F(\mathbb{N})$ for a $(2, 3)$-automatic function.

Since the degree of the reduced kernel graph is equal to 3, Lemma 4.4.13 does not apply. Let $P(x)$ be the reduced substitution polynomial then

$$P(1) = \begin{pmatrix} 0 & 1 & 2 \\ 0 & 1 & 1 \\ 1 & 0 & 0 \end{pmatrix}.$$

Since M is 3-automatic it is also 9-automatic and the reduced 9-substitution polynomial is given by $Q(x) = P(x)P(x^3)$. We obtain

$$Q(1) = P(1)^2 = \begin{pmatrix} 2 & 1 & 1 \\ 1 & 1 & 1 \\ 0 & 1 & 2 \end{pmatrix}.$$

Since the sum of the entries of $Q(1)$ in row i gives the degree of the kernel element $\underline{f}_i \in \mathcal{G}_9^*(\underline{f})$, we see that the degree of the 9-kernel graph of $\underline{f}$ is equal to 4. By Lemma 4.4.13 we can construct a $(4, 9)$-automatic function $F : \mathbb{N} \to M$ such that $F(\mathbb{N}) = M$. If F is $(4, 9)$-automatic, then it is also $(2, 3)$ automatic which proves our assertion.

The reduced 9-kernel graph of M is shown in Figure 4.16.

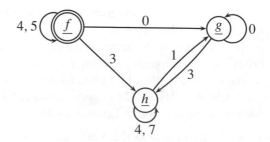

Figure 4.16. The reduced 9-kernel graph of M in Figure 4.15.

Moreover, for any $p \geq 2$ there exists a $(p, 3)$-automatic function such that M is the image of the function.

The above examples indicate that the existence of a surjective function from $\mathbb{N}$ onto an automatic set $M \subset \mathbb{N}$ is closely related to the behavior of the powers of the matrix $P_M(1)$. This behavior is influenced by the eigenvalues of the matrix $P(1)$. Note further that $P(1)$ is a non-negative matrix, i.e., all entries of $P(1)$ are greater than or equal to zero.

Remark. If A is an $N \times N$ matrix with entries $a_{ij} \geq 0$, then there exists a non-negative eigenvalue λ_+ such that $\lambda_+ \geq |\lambda|$ for all other eigenvalues λ of A, see e.g. [84]. The eigenvalue is called *leading eigenvalue* of A.

As a first result on the existence of a surjective (p, q)-automatic function $F : \mathbb{N} \to M$ we state

Theorem 4.4.22. *Let M_q be a q-automatic subset, $P_M(x)$ its reduced q-substitution polynomial and let λ_+ be the leading eigenvalue of $P_M(1)$. If $p \geq 2$ and $p > \lambda_+$, then there exists a (p, q)-automatic function $F : \mathbb{N} \to \mathbb{N}$ such that $F(\mathbb{N}) = M_q$.*

Proof. Let $\epsilon > 0$ be such that $\epsilon + \lambda_+ < p$. Since λ_+ is the leading eigenvalue there exists a constant $r > 0$ such that the entries $(p_{ij}^n)_{i,j=1,\ldots,N}$ of the nth power $P_M(1)^n$ satisfy

$$0 \leq p_{ij}^n < r(\epsilon + \lambda_+)^n$$

for all $n \in \mathbb{N}$, see [84]. Thus there exists an $n_0 \in \mathbb{N}$ such that

$$\sum_{j=1}^{N} p_{ij}^{n_0} \leq N(\epsilon + \lambda_+)^{n_0} < p^{n_0}$$

holds for all $i = 1, \ldots, N$. Since the sum of the entries in the i-th row of $P_M(1)^{n_0}$ equals the degree of $\underline{f}_i$ in the reduced $P_M^{n_0}$-kernel graph, we obtain that the degree of the reduced p^{n_0}-kernel graph is less than p^{n_0}. By Lemma 4.4.13, there exists a (p^{n_0}, q^{n_0})-automatic map F such that $F(\mathbb{N}) = M_q$. This proves the assertion. $\square$

The above theorem is useful if the leading eigenvalue is not a natural number. Example 1. from above shows that it is not necessarily true that $\lambda_+ = p$ implies the existence of a (p, q)-automatic function F with $F(\mathbb{N}) = M$. Example 2. provides an example where the leading eigenvalue is less than 2. In fact the leading eigenvalue is approximately 1.8019.

The following examples show that a surjective function may exist even if the leading eigenvalue of $P_M(1)$ is an integer.

Examples.

1. Figure 4.17 shows an example of a reduced 3-kernel graph of a 3-automatic subset M. It is plain to see the $P(1) = 2$, therefore the leading eigenvalue is equal to 2. Due to Lemma 4.4.13, there exists a $(2, 3)$-automatic function F such that $F(\mathbb{N}) = M$.

$$0, 2 \; \underline{f}$$

Figure 4.17. 3-automatic subset $M \subset \mathbb{N}$ that is of the form $M = F(\mathbb{N})$ for a $(2, 3)$-automatic function.

2. Figure 4.18 presents an example of a reduced 3-kernel graph that generates a 3-automatic set M such that the leading eigenvalue of $P(1)$ is equal to 2. Note that $\underline{f}(0) = 1$ and $\underline{g}(0) = 1$.

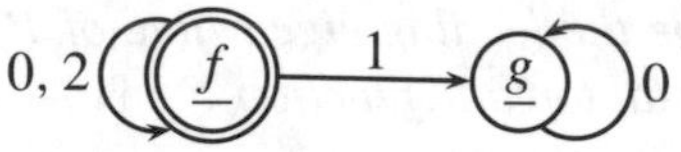

$$0, 2 \; \underline{f} \xrightarrow{\ 1\ } \underline{g} \; 0$$

Figure 4.18. 3-automatic subset $M \subset \mathbb{N}$ with degree of the kernel graph equal to 3 and that is of the form $M = F(\mathbb{N})$ for a $(2, 3)$-automatic function.

One computes that the degree of $\underline{f}$ in the p^n-kernel graph is given by

$$\deg(\underline{f} \in \mathcal{G}^*_{p^n}) = 2^{n+1} - 1.$$

Therefore $\deg(\underline{f} \in \mathcal{G}^*_{p^n}) > 2^n$ for all $n \geq 1$ and Lemma 4.4.13 does not apply. On the other hand, the 3-counting function $C_3(M)$ is also given by

$$C_3(M) = 2^{n+1} - 1,$$

which might induce the conjecture that there exists a $(2, 3)$-automatic function F with $F(\mathbb{N}) = M$. Indeed, it is possible to construct a $(2, 3)$-automatic function $F : \mathbb{N} \to \mathbb{N}$ such that $F(\mathbb{N}) = M$. We outline the construction of this function. At first, one observes that M can be partitioned into two non-empty subsets M_0, M_1 such that $M = M_0 \cup M_1$. These are

$$M_0 = \text{supp}(\underline{f}) \quad \text{and} \quad M_1 = \text{supp}(\underline{g}).$$

Then there exists a surjective $(2, 3)$-automatic function $F_0 : \mathbb{N} \to M_0$ (the reduced kernel graph for χ_{M_0} is shown in Figure 4.17). Figure 4.19 shows the reduced $(2, 3)$-transition graph of a $(2, 3)$-automaton that generates a surjective function $F_1 : \mathbb{N} \setminus \{0\} \to M_1$, where $\omega(\underline{f}) = 0$ and $\omega(\underline{g}) = 1$.

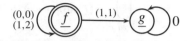

Figure 4.19. Reduced $(2, 3)$-transition graph that generates $F_1 : \mathbb{N} \setminus \{0\} \to M_1$.

Lemma 4.1.8 now states that the function $F : \mathbb{N} \to M$ defined by

$$F(n) = \begin{cases} F_0(k) & \text{if } n = 2k \\ F_1(k) & \text{if } n = 2k - 1 \end{cases}$$

is $(2, 3)$-automatic and, due to our construction, F is surjective.

As we have already seen, if the leading eigenvalue of $P(1)$ is equal to a natural number greater than or equal to 2 then the existence of a surjective map F depends on the finer structure of the reduced kernel graph. Before we study this dependence we introduce the notion of recurrence.

Definition 4.4.23. Let $\chi \in \Sigma(\mathbb{N}, \mathbb{B})$ be p-automatic. If there exists a $k \in \mathbb{N}$ and $v_0, \ldots, v_k \in \{0, \ldots, p-1\}$ such that

$$\partial_{v_0} \circ \cdots \circ \partial_{v_k}(\chi) = \chi,$$

then χ is called p-recurrent.

The importance of the recurrent sequences is demonstrated by the following theorem.

Theorem 4.4.24. *Let $M \subset \mathbb{N}$ be q-automatic such that $\chi = \chi_M$ is not q-recurrent. Then M is the image of a (p, q)-automatic function $F : \mathbb{N} \to M$ if and only there exist pairwise disjoint q-automatic non-empty sets $M_j \subset \mathbb{N}$, $j = 1, \ldots, l$ and (p, q)-automatic functions $F_j : \mathbb{N} \to M_j$ such that*

$$F_j(\mathbb{N}) = M_j$$
$$\bigcup_{j=1}^{l} M_j = M.$$

Proof. Since χ is q-automatic and non recurrent, there exists a $k \in \mathbb{N}$ such that the sequence

$$\partial_{v_k} \circ \cdots \circ \partial_{v_0}(\chi)$$

for $v_0, \ldots, v_k \in \{0, \ldots, q-1\}$ is either the sequence $\underline{0}$, the zero sequence, or the sequence is q-recurrent. For $Q = q^{k+1}$ and $V = \{0, \ldots, Q-1\}$ these sequences can be written as $\chi_J := \partial_J(\chi)$, where $J \in V$. Then M is given by

$$M = \bigcup_{J \mid \chi_J \neq \underline{0}} (Q \text{ supp}(\chi_J) + J),$$

where $Q \text{ supp}(\chi_J) + J = \{Qx + J \mid x \in \text{supp}(\chi_J)\} = M_J$. The sets M_J are mutually disjoint and each M_J is q-automatic.

Let us now assume that there exist a (p, q)-automatic function $F : \mathbb{N} \to M$ such that $F(\mathbb{N}) = M$. It remains to show that there exist (p, q)-automatic functions $F_J : \mathbb{N} \to M_J$ with $F_J(\mathbb{N}) = M_J$ for all $J \in \{0, \ldots, Q-1\}$ with $M_J \neq \emptyset$. To this end we construct (q, q)-automatic functions $G_J : M \to M_J$ such that $G_J(M) = M_J$. The functions F_J are then given by $F_J = G_J \circ F$. Let $m_J \in M_J$ be fixed for every J with $M_J \neq \emptyset$. The functions $G_J : M \to M_J$ defined by

$$G_J(m) = \begin{cases} m & \text{if } m \in M_J \\ m_J & \text{otherwise} \end{cases}$$

are (q, q)-automatic and satisfy $G_J(M) = M_J$.

The other assertion, namely that $M = F(\mathbb{N})$ for a (p, q)-automatic function F if M is the union of finitely many mutually disjoint sets $M_1, \ldots, M_l$ with (p, q)-automatic functions $F_j : \mathbb{N} \to M_j$, $j = 1, \ldots l$, and $F_j(\mathbb{N}) = M_j$ is a direct consequence of Lemma 4.1.8. □

Thus, non-recurrent sequences can be considered as being composed of recurrent sequences. In order to study recurrent sequences, we make deliberate use of the theory of non-negative matrices, see, e.g., [84] or [151].

We begin with some standard definitions. A matrix $A = (a_{ij}) \in \mathbb{R}^{n \times n}$ is called non-negative if $a_{ij} \geq 0$ for all $i, j = 1, \ldots, n$. The matrix A is called positive if $a_{ij} > 0$ for all $i, j = 1, \ldots, n$.

Definition 4.4.25. A non-negative matrix $A \in \mathbb{R}^{n \times n}$ is *primitive* if there exists an $n_0 \in \mathbb{N}$ such that A^{n_0} is positive.

A non-negative matrix $A = (a_{ij}) \in \mathbb{R}^{n \times n}$ is *irreducible* if for every pair $(i, j) \in \{1, \ldots, n\}$ there exists an $n \in \mathbb{N}$ such that (i, j)-th entry of A^n is positive.

Remarks.
1. If a non-negative matrix A is reducible, then there exist a permutation matrix P such that the matrix $A' = P^{-1}AP$ is of upper triangular block form:

$$A' = \begin{pmatrix} A_1 & \star & \cdots & \star \\ 0 & A_2 & \ddots & \vdots \\ \vdots & \ddots & \ddots & \star \\ 0 & \cdots & 0 & A_k \end{pmatrix}$$

where the A_i, $i = 1, \ldots, k$, are either irreducible square matrices or the zero matrix.

2. If a non-negative matrix A is irreducible, then either A is primitive or there exists a permutation matrix P and a $\pi \in \mathbb{N}$, $\pi \geq 2$, such $A' = P^{-1}AP$ is of the form

$$
A' = \begin{pmatrix}
0 & D_0 & 0 & \cdots & 0 \\
0 & 0 & D_1 & \cdots & 0 \\
\vdots & \vdots & \vdots & \ddots & \vdots \\
0 & 0 & 0 & \cdots & D_{\pi-2} \\
D_{\pi-1} & 0 & 0 & \cdots & 0
\end{pmatrix}.
$$

Moreover $(A')^\pi$ is of block diagonal form

$$
(A')^\pi = \begin{pmatrix}
A_0 & 0 & \cdots & 0 \\
0 & A_1 & & \vdots \\
\vdots & & \ddots & 0 \\
0 & \cdots & 0 & A_{\pi-1}
\end{pmatrix},
$$

where the square matrices A_i, $i = 0, \ldots, \pi - 1$, are primitive and given by $A_i = D_i \ldots D_{\pi-1} D_0 \ldots D_{i-1}$.

If $A \in \mathbb{R}^{n \times n}$ is given and if λ_0 is an eigenvalue of A, then the order of the zero of the characteristic polynomial of A, i.e., $\det(\lambda \operatorname{id} - A)$, is called the *algebraic multiplicity* of λ_0. The complex dimension of the kernel of the matrix $\lambda \operatorname{id} - A$ is called the *geometric multiplicity* of λ_0.

The main result of the Perron–Frobenius theory is the next theorem.

Theorem 4.4.26 (Perron–Frobenius Theorem). *If $A \in \mathbb{R}^{n \times n}$, $A \neq 0$, is a non-negative irreducible matrix, then A has a positive eigenvector $x \in \mathbb{R}^n$ with eigenvalue $\lambda_+ > 0$ that has algebraic and geometric multiplicity equal to one. Furthermore, any positive eigenvector of A is a multiple of x, and if μ is another eigenvalue of A, then $|\mu| \leq \lambda_+$.*

Proofs may be found in the above mentioned books. Note that if A is a non-negative matrix in upper triangular block form, then the leading eigenvalue λ_+ of A is the maximum of the leading eigenvalues of the block matrices A_i.

Remarks.

1. If $A \in \mathbb{R}^{n \times n}$ is a non-negative and irreducible matrix, then the leading eigenvalue λ_+ of A satisfies

$$
r_- \leq \lambda_+ \leq r_+,
$$

where r_- and r_+ denote the smallest and the largest row sum of A, respectively.

2. If $A \neq 0$ is a non-negative irreducible matrix such that A is of the form

$$A = \begin{pmatrix} 0 & D_0 & 0 & \dots & 0 \\ 0 & 0 & D_1 & \dots & 0 \\ \vdots & \vdots & \vdots & \ddots & \vdots \\ 0 & 0 & 0 & \dots & D_{\pi-2} \\ D_{\pi-1} & 0 & 0 & \dots & 0 \end{pmatrix},$$

and if λ is an eigenvalue of A and $\mu \in \mathbb{C}$ a π-th root of unity, then $\mu\lambda$ is also an eigenvalue of A.

3. Let A be as above. Since A^π is of block diagonal form

$$A^\pi = \begin{pmatrix} A_0 & 0 & \dots & 0 \\ 0 & A_1 & & \vdots \\ \vdots & & \ddots & 0 \\ 0 & \dots & 0 & A_{\pi-1} \end{pmatrix}$$

with primitive matrices A_i, each of the matrices A_i has the leading eigenvalue λ_+^π, where λ_+ is the leading eigenvalue of A.

If $M \subset \mathbb{N}$, $M \neq \emptyset$, is a p-automatic set and if $P_M(x)$ denotes the reduced substitution polynomial then $P_M(1)$ is clearly a non-negative matrix. Remark 1, p. 136, can be stated as follows: There exists a numbering of the kernel elements of χ_M such that $P_M(1)$ has an upper triangular block structure. The irreducible matrices $A_{ii} \neq 0$ correspond to strongly connected components of the reduced kernel graph of χ_M. A non-empty subset $C \subset \ker_p(\chi_M)$ is called strongly connected if for every $\underline{g}$, $\underline{h} \in G$ there exists a path in G from $\underline{h}$ to $\underline{g}$ and vice versa. Furthermore C is maximal with this property (w.r.t. inclusion).

If C_1, $C_2 \subset \ker_p(\chi_M)$ are two different strongly connected components of the reduced kernel graph of χ, then C_1 is larger than C_2, denoted by $C_1 \succ C_2$, if there exist $\underline{g} \in C_1$ and $\underline{h} \in C_2$ such that $\underline{h} \in \ker_p(\underline{g})$ or, equivalently, there exists a path from $\underline{g}$ to $\underline{h}$ in the reduced kernel graph of χ_M. This clearly defines a partial order on the set of strongly connected components. If χ_M is p-recurrent, the strongly connected component containing χ_M is the maximal element.

Note further that for every strongly connected component C of the kernel graph of χ there exists a corresponding square matrix in the upper triangular block form of the matrix $P(1)$. This matrix is denoted by $A(C)$.

Theorem 4.4.27. *Let $M \subset \mathbb{N}$ be q-automatic and let χ_M be q-recurrent. Let $P(x)$ be the reduced substitution polynomial and let $\lambda_+ = p \geq 2$, $p \in \mathbb{N}$, be the leading eigenvalue of $P(1)$.*

If there exist two strongly connected components C_1, $C_2 \subset \ker_q(\chi_M)$ such that $C_1 \prec C_2$ and such that the leading eigenvalues of $A(C_1)$ and $A(C_2)$ are both equal to p, then there exists no (p, q)-automatic function $F : \mathbb{N} \to \mathbb{N}$ such that $F(\mathbb{N}) = M$.

Proof. We assume that $P(1)$ is given in upper triangular block structure. Since the matrices A_i, $i = 1, \ldots, k$, along the diagonal are either irreducible or zero, there exists an $N_0 \in \mathbb{N}$ such that either $A_i^N = 0$ or A_i^N is of the form

$$
\begin{pmatrix}
A_{i0} & 0 & \cdots & 0 \\
0 & A_{i1} & & \vdots \\
\vdots & & \ddots & 0 \\
0 & \cdots & 0 & A_{i\pi_i - 1}
\end{pmatrix}
$$

if A_i is irreducible. Since the matrices $A_{i,j}$ are primitive there exists an $N_1 \in \mathbb{N}$ such that the $A_{i,j}^{N_1}$ are positive matrices. This yields that $P(1)^{N_0 N_1}$ is of upper triangular block structure and a block along the diagonal is either a positive square matrix or the zero matrix.

Let A_{i_1} and A_{i_2} be the matrices corresponding to the strongly connected components C_1 and C_2. Since $C_1 \succ C_2$, the triangular block structure of $P(1)$ implies that $P(1)$ contains the submatrix

$$
\begin{pmatrix}
A_{i_1} & \star & \cdots & \star \\
0 & A_{i_1+1} & \ddots & \vdots \\
\vdots & \ddots & \ddots & \star \\
0 & \cdots & 0 & A_{i_2}
\end{pmatrix}.
$$

Now the results on the structure of $P(1)^{N_0 N_1}$ and the fact that $C_1 \succ C_2$ imply the existence of an N_2 such that $P(1)^{N_0 N_1 N_2}$ contains the submatrix

$$
\begin{pmatrix}
A_{i_1 j_1}^{N_2} & \cdots & B \\
\vdots & \ddots & \vdots \\
0 & \cdots & A_{i_2 j_2}^{N_2}
\end{pmatrix},
$$

where $A_{i_1 j_1}$, $j_1 \in \{0, \ldots, \pi_1 - 1\}$, and $A_{i_2 j_2}$, $j_2 \in \{0, \ldots, \pi_2 - 1\}$, are primitive and positive matrices and where the matrix $B = (b_{st})$ with $(s, t) \in I \times J$ for certain sets $I, J \subset \{1, \ldots, |\ker_q(\chi)|\}$ is positive. Due to the assumption, A_{i_1} and A_{i_2} both have the leading eigenvalue p. Due to the primitivity of $A_{i_1 j_1}$ and $A_{i_2 j_2}$ it follows that there exists a constant $c > 0$ such that

$$
A_{i_1 j_1}^{N N_0 N_1 N_2} > c p^{N N_0 N_1 N_2}
$$

$$
A_{i_2 j_2}^{N N_0 N_1 N_2} > c p^{N N_0 N_1 N_2}
$$

holds for all $N \in \mathbb{N}$. If $B_N = (b_{st}^N)$ denotes the submatrix of $P(1)^{N N_0 N_1 N_2}$ defined by the entries at position $(s, t) \in I \times J$, then an induction argument shows that

$$
b_{st}^N \geq 2cp^{(N-1)N_0 N_1 N_2} + (N-2)c^2 p^{(N-1)N_0 N_1 N_2}
$$

for all $N \in \mathbb{N}$ and all $(s, t) \in I \times J$.

Thus there exists a sequence $\underline{h} \in \ker_q(\chi_M)$ such that

$$C_q(\text{supp}(\underline{h}))(N N_0 N_1 N_2) \geq 2cp^{(N-1)N_0 N_1 N_2} + (N-2)c^2 p^{(N-1)N_0 N_1 N_2}$$

for all $N \in \mathbb{N}$. Due to Lemma 4.4.16, a (p, q)-automatic function $F : \mathbb{N} \to \text{supp}(\underline{h})$ with $F(\mathbb{N}) = \text{supp}(\underline{h})$ does not exist. Since $\underline{h} \in \ker_q(\chi_M)$ the same conclusion holds for the set M.   □

In short the above theorem says, that if there exists a bad connection between two strongly components of the reduced q-kernel graph of χ_M, then M is not an image of a (p, q)-automatic function. If no such bad connection exist, then the question on the existence of a (p, q)-automatic function $G\mathbb{N} \to M$ with $G(\mathbb{N}) = M$ is still open.

We conclude this section with a further result which can be derived from our approach to compute $C_p(M)$.

Lemma 4.4.28. *Let $M \subset \mathbb{N}$, $M \neq \emptyset$, be a p-automatic subset such that χ_M is p-recurrent and let λ_+ be the leading eigenvalue of $P^{\text{red}}(1)$. There exists a $c > 0$ and an $n_0 \in \mathbb{N}$ such that*

$$C_p(M)(n) \geq c\lambda_+^n$$

holds for all $n \geq n_0$.

Proof. Since M is recurrent and non-empty, the leading eigenvalue λ_+ of $P^{\text{red}}(1)$ is positive. Furthermore, we assume that $P^{\text{red}}(1)$ is in upper triangular block form. Then there exists a submatrix A_i along the diagonal of $P^{\text{red}}(1)$ such that A_i is irreducible and has leading eigenvalue λ_+. Due to the irreducibility there exists a π and an n_0 such that $A_i^{\pi n_0}$ is of block diagonal form, where the square matrices $A_{i1}, \ldots, A_{i\pi}$ along the diagonal are positive and with leading eigenvalue $\lambda_+^{\pi n_0}$. We can therefore conclude that there exists a $c > 0$ such that

$$A_{ij}^{\pi n_0 n} > c\lambda_+^{\pi n_0 n}$$

holds for all $n \in \mathbb{N}$, $n \neq 0$, and all $j = 1, \ldots, \pi$. This shows that there exists an $\underline{h} \in \ker_p(\chi_M)$ such that

$$C_p(\text{supp}(\underline{h}))(\pi n_0 n) > c\lambda_+^{\pi n_0 n}$$

for all $n \in \mathbb{N}$, $n \geq 1$. This leads to

$$C_p(\text{supp}(\underline{h}))(\pi n_0 + n) > \frac{c}{\lambda_+^{\pi n_0 - 1}}\lambda^{\pi n_0 + n}$$

for all $n \geq 1$. Since $\underline{h} \in \ker_p(\chi_M)$, it follows that there exist $m \in \mathbb{N}$ and $l \in \{0, \ldots, p^m - 1\}$ such that

$$p^m \text{supp}(\underline{h}) + l \subset M.$$

This yields the existence of a $c' > 0$ such that

$$C_p(\pi n_0 + n + l + 1) \geq c' \lambda_+^{\pi n_0 + n + 1 + l}$$

holds for all $n \geq 1$. This proves the assertion. □

The case χ_M being non-recurrent is similar.

Corollary 4.4.29. *Let $M \subset \mathbb{N}$ be p-automatic. If λ_+ denotes the leading eigenvalue of $P^{\mathrm{red}}(1)$, then there exist $c > 0$ and $n_0 \in \mathbb{N}$ such that*

$$C_p(M)(n) \geq \begin{cases} c\lambda_+^n & \text{if } \lambda_+ > 0 \\ c & \text{if } \lambda_+ = 0 \end{cases}$$

holds for all $n \geq n_0$.

Proof. If $\lambda_+ > 0$, then the assertion is a consequence of Lemma 4.4.28. If $\lambda_+ = 0$, then the upper triangular block form of $P^{\mathrm{red}}(1)$ has only zero matrices along the diagonal. In other words, the reduced kernel graph has no strongly connected component. This means that M contains only finitely many points. This proves the second assertion. □

In combination with the Jordan form of the matrix $P^{\mathrm{red}}(1)$ Lemma 4.4.28 and Corollary 4.4.29 one obtains a rather good description of the growth of $C_p(M)(n)$.

Theorem 4.4.30. *Let $M \subset \mathbb{N}$ be p-automatic and unbounded, and let λ_+ be the leading eigenvalue of $P^{\mathrm{red}}(1)$. There exist $\alpha \in \mathbb{N}$, $k \in \mathbb{N}$ and $s \in [0, 1]$ such that the limit*

$$\lim_{n \to \infty} \frac{C_p(M)(\alpha n)}{(\pi n)^k p^{\alpha n s}}$$

exists and is positive.

Proof. Let N denote the cardinality of the reduced p-kernel of χ_M. Let $P^{\mathrm{red}}(1)$ be in upper triangular block structure. Then there exists an $\alpha \in \mathbb{N}$ such that $P^{\mathrm{red}}(1)^\alpha$ has only primitive or zero matrices along its diagonal. Moreover, λ_+^α is the leading eigenvalue and all other eigenvalues μ of $P^{\mathrm{red}}(1)^\alpha$ satisfy $|\mu| < \lambda_+^\alpha$.

We can therefore conclude that there exists a matrix $U \in \mathbb{C}^{N \times N}$ such that

$$P^{\mathrm{red}}(1)^\alpha = U \begin{pmatrix} \Lambda_* & 0 & 0 \\ 0 & \Lambda & 0 \\ 0 & 0 & \tilde{L} \end{pmatrix} U^{-1},$$

where Λ is a diagonal matrix with entries λ_+^α along the diagonal, Λ_* has entries λ_+^α along the diagonal and entries 1 at the positions $(j, j+1)$, the matrix $\tilde{L}$ is an upper triangular matrix and the entries along the diagonal have absolute value less then λ_+^α.

Now choose $s = \frac{\log \lambda_+}{\log p}$. Since λ_+ is an eigenvalue of an irreducible matrix which corresponds to a strongly connected component of the p-kernel graph of χ, it follows that $1 \leq \lambda_+ \leq p$ and therefore $s \in [0, 1]$. Note that $p^s = \lambda_+$. Then we have

$$
\frac{1}{p^{\alpha n s}} P^{\mathrm{red}}(1)^{\alpha n} = U \begin{pmatrix} \frac{1}{p^{\alpha n s}} \Lambda_*^n & 0 & 0 \\ 0 & \frac{1}{p^{\alpha n s}} \Lambda^n & 0 \\ 0 & 0 & \frac{1}{p^{\alpha n s}} \tilde{L}^{\alpha n} \end{pmatrix} U^{-1}
$$

for all $n \in \mathbb{N}, n \neq 0$. If Λ_* is a $k \times k$-matrix, where $k \geq 2$, then the nth power of Λ_* is given by

$$
\Lambda_*^n = \begin{pmatrix} \lambda_+^{\alpha n} & \binom{n}{1} \lambda_+^{\alpha n - 1} & \cdots & \binom{n}{k-1} \lambda_+^{\alpha n - (k-1)} \\ & \ddots & \ddots & \vdots \\ & & \lambda_+^{\alpha n} & \binom{n}{1} \lambda_+^{\alpha n - 1} \\ & & & \lambda_+^{\alpha n} \end{pmatrix}.
$$

Therefore

$$
\frac{C_p(M)(\alpha n)}{p^{\alpha n s}}
$$

$$
= (1, 0, \ldots, 0) U \begin{pmatrix} \frac{1}{p^{\alpha n s}} \Lambda_*^n & 0 & 0 \\ 0 & \frac{1}{p^{\alpha n s}} \Lambda^n & 0 \\ 0 & 0 & \frac{1}{p^{\alpha n s}} \tilde{L}^{\alpha n} \end{pmatrix} U^{-1} (\underline{f}_1(0), \underline{f}_2(0), \ldots, \underline{f}_N(0))^T
$$

is of the form

$$
\frac{C_p(M)(\alpha n)}{p^{\alpha n s}} = \sum_{j=0}^{k-1} \alpha_j \binom{n}{j} \frac{1}{\lambda_+^j} + o(n),
$$

where $\lim_{n \to \infty} o(n) = 0$ and $\alpha_j \in \mathbb{R}$. Due to Lemma 4.4.29, we obtain

$$
0 < c < \frac{C_p(M)(\alpha n)}{p^{\pi n s}} = \sum_{j=0}^{k-1} \alpha_j \binom{n}{j} \frac{1}{\lambda_+^j} + o(n).
$$

This shows that at least one of the coefficients α_j is different from zero. Let $j_0 = \max\{j \mid \alpha_j \neq 0, \, j = 0, \ldots, k-1\}$, then clearly $\alpha_{j_0} > 0$. This concludes the proof, since

$$
\lim_{n \to \infty} \frac{1}{(\pi n)^k} \alpha_k \binom{n}{k} \frac{1}{\lambda_+^k} = \frac{\alpha_k}{\pi^k \lambda_+^k k!} > 0
$$

is the desired limit. $\square$

As an application we show that the prime numbers are not automatic. It is a well-known fact that the number of prime numbers in the interval $[0, n]$ is given by

$O(n/\log n)$, i.e., there exist constants $0 < c_1 < c_2$ such that

$$c_1 \frac{x}{\log x} \leq |\{p \mid p \text{ is prime}, \ p \in \mathbb{N} \ p \leq x\}| \leq c_2 \frac{x}{\log x}$$

holds for all $x > 0$. Actually there exist stronger estimates. However, the above estimate is sufficient for our purpose.

Now assume that the set of prime numbers $\boldsymbol{P}$ is p-automatic. Then there exist π, $k \in \mathbb{N}$ and $s \in [0, 1]$ such that the limit

$$\lim_{n \to \infty} \frac{C_p(\boldsymbol{P})(\pi n)}{p^{\pi n s} (\pi n)^k}$$

exists and is positive. The asymptotic formula for $\pi(n)$ yields

$$C_p(\boldsymbol{P})(\pi n) = O\left(\frac{p^{\pi n}}{\pi n \log p} \right)$$

which in turn gives

$$\frac{C_p(\boldsymbol{P})(\pi n)}{p^{\pi n s} (\pi n)^k} = O\left(\frac{p^{\pi n}}{(\pi n)^{k+1} p^{\pi n s} \log p} \right).$$

This shows that $s = 1$, in order to guarantee the existence of a limit. However, for every choice of $k \in \mathbb{N}$, the limit is zero. Therefore it follows that the set of prime numbers is not p-automatic for all $p \geq 2$.

4.5 Cellular automata and automatic maps

In this section we shall show how cellular automata may be used to define $(H_1 \times H_2)$-automatic maps $G : \Gamma_1 \to \Gamma_2$. Surprisingly, the conditions on a cellular automaton to induce an automatic map are very weak.

We begin with a definition of a cellular automaton. Let $\mathcal{A}_1$, $\mathcal{A}_2$ be finite sets. As we have mentioned in Chapter 1 we can define a metric Δ_1 on $\Sigma(\mathbb{Z}, \mathcal{A}_1)$ and a metric Δ_2 on $\Sigma(\mathbb{Z}, \mathcal{A}_2)$ in such a way that both sequence spaces become complete metric spaces. It is therefore meaningful to consider continuous functions between the sequence spaces. Among the continuous maps there is a special class of continuous maps, the class of cellular automata.

Definition 4.5.1. Let $\Psi : (\Sigma(\mathbb{Z}, \mathcal{A}_1), \Delta_1) \to (\Sigma(\mathbb{Z}, \mathcal{A}_2), \Delta_2)$ be a continuous map. Ψ is called a *cellular automaton* if

$$\Psi((T_x)_*(\underline{f})) = (T_x)_*(\Psi(\underline{f}))$$

holds for all $\underline{f} \in \Sigma(\mathbb{Z}, \mathcal{A}_1)$.

As usual, we denote an element $\underline{f} \in \Sigma(\mathbb{Z}, \mathcal{A}_1)$ by

$$\underline{f} = \bigoplus_{j \in \mathbb{Z}} f_j x^j$$

and $(T_x)_*(\oplus f_j x^j) = \oplus f_j x^{j+1} = x\underline{f}$ is called shift, see Chapter 1.

The following theorem is due to Hedlund, [97]. It gives a complete characterization of cellular automata.

Theorem 4.5.2. *A continuous map* $\Psi : (\Sigma(\mathbb{Z}, \mathcal{A}_1), d_1) \to (\Sigma(\mathbb{Z}, \mathcal{A}_2), d_2)$ *is a cellular automaton if and only if there exist a* $k \in \mathbb{N}$ *and a function* $\phi : \mathcal{A}_1^{2k+1} \to \mathcal{A}_2$ *such that*

$$\Psi(\underline{f})(x^j) = \phi(\underline{f}(x^{j-k}), \dots, \underline{f}(x^j), \dots, \underline{f}(x^{j+k}))$$

holds for all $\underline{f} \in \Sigma(\mathbb{Z}, \mathcal{A}_1)$ *and for all* $j \in \mathbb{Z}$.

Remarks.

1. The function $\phi : \mathcal{A}_1^{2k+1} \to \mathcal{A}_2$ is called the generating function of the cellular automaton Ψ.

2. Let $H : \Gamma \to \Gamma$ and let V_c be a complete digit set of H. If $\gamma \in \Gamma$, then

$$\underline{p}_\gamma = \bigoplus_{j=0}^{k} v_j x^j,$$

where $\gamma = v_0 H(v_1) \dots H^{\circ k}(v_k)$ is a polynomial in $\Sigma_c(\mathbb{Z}, V_c)$. Note that $\underline{p}_e$ is the polynomial e. $\underline{p}_\gamma$ is called the γ-polynomial.

If $\underline{q} \in \Sigma_c(\mathbb{Z}, V_c)$ is a polynomial, i.e., the set $\{j \mid \underline{q}(j) \neq e\}$ is finite, then

$$\mathfrak{p}(\underline{q}) = \underline{q}(0) H(\underline{q}(1)) H^{\circ 2}(\underline{q}(2)) \dots H^{\circ k}(\underline{q}(k)) \dots, \qquad (4.1)$$

is a well-defined element of Γ. Moreover, one has

$$\mathfrak{p}(\underline{p}_\gamma) = \gamma$$

for all $\gamma \in \Gamma$. In other words, $\mathfrak{p}$ is the inverse of the map $\gamma \mapsto \underline{p}_\gamma$. Note that $\underline{p}_{\mathfrak{p}(\underline{q})} = \underline{q}$ if and only if $\underline{q}$ truly is a polynomial. i.e. $\operatorname{supp}(\underline{q}) \subset \mathbb{N}$.

We now set $\mathcal{A}_1 = V_c$ and $\mathcal{A}_2 = W_c$, where V_c and W_c are complete digit sets for the expanding maps $H_1 : \Gamma_1 \to \Gamma_1$ and $H_2 : \Gamma_2 \to \Gamma_2$, respectively. Furthermore, the neutral elements $e_1 \in V_c$ and $e_2 \in W_c$ play the role of the empty symbol in $\mathcal{A}_1$ and $\mathcal{A}_2$.

Definition 4.5.3. Let $\Psi : \Sigma(\mathbb{Z}, V_c) \to \Sigma(\mathbb{Z}, W_c)$ be a cellular automaton. The cellular automaton Ψ *preserves polynomials* if

$$\Psi(\Sigma_c(\mathbb{Z}, V_c)) \subset \Sigma_c(\mathbb{Z}, W_c).$$

Remarks.

1. A cellular automaton $\Psi : \Sigma(\mathbb{Z}, V_c) \to \Sigma(\mathbb{Z}, W_c)$ preserves polynomials if and only if

$$\phi(e_1, \dots, e_1) = e_2$$

   for the generating function $\phi : V_c^{2k+1} \to W_c$ of Ψ.

2. A special class of cellular automata which preserve polynomials is given by the *left dependent* cellular automata. A cellular automaton with generating function $\phi : V_c^{2k+1} \to W_c$ is left dependent if there exists a map $\tilde{\phi} : V_c^{k+1} \to W_c$ such that

$$\phi(a_{-k}, \dots, a_0, a_1, \dots, a_k) = \tilde{\phi}(a_{-k}, \dots, a_{-1}, a_0)$$

   holds for all $(a_{-k}, \dots, a_k) \in V_c^{2k+1}$.

If $\Psi : \Sigma(\mathbb{Z}, V_c) \to \Sigma(\mathbb{Z}, W_c)$ is a cellular automaton that preserves polynomials, then there exists an *induced map* $G_\Psi : \Gamma_1 \to \Gamma_2$ defined by

$$G_\Psi(\gamma_1) = \mathfrak{p} \circ \Psi(\underline{p}_{\gamma_1}),$$

where $\underline{p}_{\gamma_1} \in \Sigma_c(\mathbb{Z}, V_c)$ is the γ_1-polynomial and $\mathfrak{p} : \Sigma(\mathbb{Z}, W_c) \to \Gamma_2$ is defined as in Equation (4.1).

Theorem 4.5.4. *If $\Psi : \Sigma(\mathbb{Z}, V_c) \to \Sigma(\mathbb{Z}, W_c)$ is a cellular automaton that preserves polynomials, then the induced map $G_\Psi : \Gamma_1 \to \Gamma_2$ is $(H_1 \times H_2)$-automatic.*

Proof. Let $\chi \in \Sigma(\Gamma_1 \times \Gamma_2, \mathbb{B})$ denote the characteristic sequence of the graph of $G_\Psi : \Gamma_1 \to \Gamma_2$. The generating function of the cellular automaton Ψ is denoted by $\phi : A_1^{2k+1} \to A_2$. It is no restriction to assume that k is greater than or equal to 1.

In order to prove the finiteness of the $(V_c \times W_c)$-kernel of χ, we define auxiliary functions $\Phi_\alpha^\beta : \Gamma_1 \to \Gamma_2$, where $\alpha \in V_c^k$ and $\beta \in V_c^{k+1}$. For reasons that will be apparent in a moment we denote α by $\alpha = (a_{-k}, \dots, a_{-1})$ and β by $\beta = (b_0, b_1, \dots, b_k)$. With these settings we define

$$\Phi_\alpha^\beta : \Gamma_1 \to \Gamma_2$$

by

$$\Phi_\alpha^\beta(\gamma_1) = \mathfrak{p} \circ \Psi\Big(\bigoplus_{i=-k}^{-1} a_i x^i \oplus \bigoplus_{i=0}^{k} b_i x^i \oplus x^{k+1} \underline{p}_{\gamma_1} \Big),$$

where $\mathfrak{p} : \Sigma_c(\mathbb{Z}, W_c) \to \Gamma_2$ is defined as in Equation (4.1). Since Ψ preserves polynomials, the maps Φ_α^β are well defined for all choices of $\alpha \in V_c^k$, $\beta \in V_c^{k+1}$.

To each Φ_α^β we associate a sequence $\chi_\alpha^\beta \in \Sigma(\Gamma_1 \times \Gamma_2, \mathbb{B})$ which is defined by

$$\chi_\alpha^\beta = \bigvee_{\gamma \in \Gamma} (\beta_0 H_1(\beta_1) \dots H_1^{\circ k}(\beta_k) H_1^{\circ(k+1)}(\gamma), \Phi_\alpha^\beta(\gamma)).$$

Note that χ_α^β can be considered as the characteristic sequence of the graph of a function defined on the set $\beta_0 H_1(\beta_1) \ldots H_1^{\circ k}(\beta_k) H_1^{\circ(k+1)}(\Gamma)$.

The auxiliary maps allow us to express the characteristic function χ of the graph of G_Ψ as a finite sum of sequences of the form χ_α^β. Indeed, we have

$$\chi = \bigvee_{\beta \in V_c^k} \chi_{(e_1, \ldots, e_1)}^\beta.$$

Let K be the set of all finite sums of sequences of the form χ_α^β with $\alpha \in V_c^k, \beta \in V_c^{k+1}$. Since the addition is in the Boolean algebra $\mathbb{B}$, K is a finite set and contains χ.

In order to prove that G_Ψ is an automatic map it remains to show that K is invariant under decimations. It is clear that it suffices to show that the decimations of χ_α^β is a sum of sequences of the form $\chi_{\alpha'}^{\beta'}$. To this end, we make use of the following identity:

$$\Phi_\alpha^\beta(\gamma_1) = \phi(\alpha, \beta) H_2\left(\Phi_{(a_{-k+1}, \ldots, a_{-1}, b_0)}^{(b_1, \ldots, b_k, v_0)})(\gamma_1')\right) = \phi(\alpha, \beta) H_2\left(\Phi_{(\alpha', b_0)}^{(\beta', v_0)}(\gamma_1')\right),$$

where $\gamma_1 = v_0 H_1(\gamma_1')$ with $v_0 \in V_c$ and $(\beta', v_0) = (b_1, \ldots, b_k, v_0)$, $(\alpha', b_0) = (a_{-k+1}, \ldots, a_{-1}, b_0)$.

Using this identity we rewrite χ_α^β as

$$\chi_\alpha^\beta = \bigvee_{\substack{\gamma_1 \in \Gamma_1 \\ v_0 \in V_c}} \left(b_0 \ldots H_1^{\circ k}(b_k) H_1^{\circ(k+1)}(v_0) H_1^{\circ(k+2)}(\gamma), \phi(\alpha, \beta) H_2\left(\Phi_{(\alpha', b_0)}^{(\beta', v_0)}(\gamma_1)\right)\right).$$

Thus we see that the decimations of χ_α^β are given by

$$\partial_{(v,w)}(\chi_\alpha^\beta) = \begin{cases} \bigvee_{v_0 \in V_c} \chi_{(\alpha', b_0)}^{(\beta', v_0)} & \text{if } v = b_0 \text{ and } \phi(\alpha, \beta) = w \\ \\ \underline{0} & \text{otherwise.} \end{cases}$$

Therefore the non-trivial decimations of χ_α^β are given as a sum of sequences of the form $\chi_{\alpha'}^{\beta'}$. This shows that the set K is invariant under decimations and proves the $(H_1 \times H_2)$-automaticity of G_Ψ. $\qquad\square$

Examples.

1. The automatic function defined by Figure 4.1 is an example of an automatic function $G : \mathbb{N} \to \mathbb{N}$ that is induced by a cellular automaton. In fact, if we consider the set $\Sigma(\mathbb{Z}, \{0, 1\})$ as the set of formal Laurent series with coefficients in the field $\mathbb{F}_2 = \{0, 1\}$, then the multiplication of a Laurent series with the polynomial $1 + x$ defines a cellular automaton $\Psi : \Sigma(\mathbb{Z}, \mathbb{F}_2) \to \Sigma(\mathbb{Z}, \mathbb{F}_2)$.

If one sets $H(x^j) = x^{2j}$ and $V = \{x^0, x^1\}$ and defines the generating function $\phi : \{0, 1\}^3 \rightarrow \{0, 1\}$ as $\phi(a, b, c) = a + b + c \bmod 2$, then the proof of Theorem 4.5.4 allows to compute the reduced kernel graph of the induced map $G_\Psi : \mathbb{N} \rightarrow \mathbb{N}$. It turns out that the reduced kernel graph of G_Ψ is the same as is shown in Figure 4.1.

Note further that $G_\Psi(\mathbb{N})$ is the support of the Thue–Morse sequence.

2. As we have seen in the previous section the map $n \mapsto 3n$ defines a $(2, 2)$-automatic function, see Figure 4.4 for the reduced kernel graph. It is a function which is not induced by a cellular automaton $\Psi : (\mathbb{Z}, \{0, 1\}) \rightarrow \Psi(\mathbb{Z}, \{0, 1\})$. Loosely speaking, this is a result of the carry overs which occur if one performs a multiplication by 3 in the binary representation of natural numbers.

Let us assume that there exists a cellular automaton $\Psi : \Sigma(\mathbb{Z}, \{0, 1\}) \rightarrow \Sigma(\mathbb{Z}, \{0, 1\})$ which induces the multiplication by 3 on the natural numbers, i.e., the graph of $G_\Psi : \mathbb{N} \rightarrow \mathbb{N}$ is equal to the graph of the function $n \mapsto 3n$. By Hedlund's Theorem 4.5.2, there exist a k and a generating map $\phi : \{0, 1\}^{2k+1} \rightarrow \{0, 1\}$.

Choose $N = 2^{\frac{2^{2n}-1}{3}}$, then N is a natural number and the $2n + 1$-st coefficient of the binary expansion of $3N$ is equal to zero. Since N is an even number the binary expansion of $N + 1$ coincides with the binary expansion of N except for the first digit. Now we have $3(N + 1) = 2^{2n+1} + 1$, i.e., the $2n + 1$-st coefficient of the binary expansion of $3(N + 1)$ is equal to one. This shows that the value of k has to be greater than $2n + 1$. Hence there exists no cellular automaton defined on $\Sigma(\mathbb{Z}, \{0, 1\})$ that induces the multiplication by 3 on $\mathbb{N}$.

The general question whether a given $(H_1 \times H_2)$-automatic function is induced by a cellular automaton is a very delicate one. By Hedlund's theorem the question is closely related to the question whether an automatic map $G : \Gamma_1 \rightarrow \Gamma_2$ has a continuous extension $\tilde{G} : \Sigma(\mathbb{Z}, V_c) \rightarrow \Sigma(\mathbb{Z}, W_c)$.

Although the proof of Theorem 4.5.4 provides an algorithm to compute the kernel graph of G_Ψ, Ψ being a cellular automaton. Performing this computation can be quite painful.

The situation is slightly improved if we consider left dependent cellular automata.

Lemma 4.5.5. *If the cellular automaton $\Psi : \Sigma(\mathbb{Z}, V_c) \rightarrow \Sigma(\mathbb{Z}, W_c)$ preserves polynomials and is left dependent, then $G_\Psi : \Gamma_1 \rightarrow \Gamma_2$ has the transducer property.*

Proof. The proof is almost a copy of the proof of Theorem 4.5.4. Due to the left dependence of Ψ is suffices to consider auxiliary maps $\Phi_\alpha : \Gamma_1 \rightarrow \Gamma_2$ with $\alpha = (a_{-k}, \ldots, a_{-1}) \in V_c^k$. These functions are defined by

$$\Phi_\alpha(\gamma_1) = \mathfrak{p} \circ \Psi\left(\bigoplus_{i=-k}^{-1} a_i x^i \oplus \underline{p}_{\gamma_1}\right).$$

The sequences $\chi_\alpha \in \Sigma(\Gamma_1 \times \Gamma_2, \mathbb{B})$ are then given by

$$\chi_\alpha = \bigvee_{\gamma_1 \in \Gamma_1} (\gamma_1, \Phi_\alpha(\gamma_1)).$$

Following the same line of arguments as above, we arrive at

$$\partial_{v,w}(\chi_\alpha) = \begin{cases} \chi_{(a_{-k},\dots,a_{-1},v)} & \text{if } w = \phi(a_{-k+1}, \dots, a_{-1}, v) \\ \underline{0} & \text{otherwise.} \end{cases}$$

This proves the transducer property. $\square$

The cellular automaton in Example 1 is a left dependent cellular automaton. Note further that the second example provides an automatic function with the transducer property that is not induced by a cellular automaton. In other words, the converse of Lemma 4.5.5 is not true.

4.6   Notes and comments

Within the framework of formal logic, automatic subsets are also called recognizable subsets. For the relations between formal logic and automatic sets we recommend [46], [47].

The kernel graph of an automatic map can also be considered as a special transducer in the sense of [77]. Besides the literature already mentioned for finite automata we add [42], [44], [59], [74], [83], and [165] for further reading on transducers.

A proof of Theorem 4.3.14 can be found in [56], see also [80] and [137].

The results on non-negative matrices can be found in, e.g., [84], [151].

For further connections between cellular automata and automatic sequences we suggest [5], [6], [11], [29], [30], [31], [44], [90], [91], [96], and [116].

Chapter 5

Algebraic properties

As already noted it is advantageous to consider an additional structure on the set $\mathcal{A}$. In order to study automatic subsets, we introduced a Boolean structure on the set $\{0, 1\}$. In this chapter we are studying automatic sequences which take their values in $\mathcal{A}$ and $\mathcal{A}$ is a commutative and associative monoid.

In the first section, we shall show that an automatic sequence always satisfies a special equation, namely a so-called N-dimensional Mahler equation (over a monoid).

In the second section, we study certain Mahler equations (over a monoid) and provide a method to solve the equation with the help of a substitution.

In the third section, we restrict our investigations to the case that Γ is a finitely generated Abelian group and $\mathcal{A}$ carries the structure of a finite field. It will turn out that every p-automatic sequence over a commutative group and with values in a finite field of characteristic p satisfies a generalized Mahler equation. We also discuss the converse of this statement.

5.1   Additional structure on $\overline{\mathcal{A}}$

In this section, we suppose that the finite set $\overline{\mathcal{A}}$ carries an algebraic structure. Throughout this section we assume that $\overline{\mathcal{A}}$ is a commutative, associative monoid with neutral element. This monoid is denoted by $\mathcal{M}$ and its neutral element is denoted by 0. The binary operation is given by $+ : \mathcal{M} \times \mathcal{M} \to \mathcal{M}$, $(a, b) \mapsto a + b$, called addition. It satisfies $a + b = b + a$ for all $a, b \in \mathcal{M}$ and $a + 0 = a$ for all $a \in \mathcal{M}$. Moreover, $a + (b + c) = (a + b) + c$ holds for all $a, b, c \in \mathcal{M}$. From now on 0 is the distinguished element of $\mathcal{M}$. In order to emphasize the addition on $\mathcal{M}$, we write sequences $\underline{f} \in \Sigma(\Gamma, \mathcal{M})$ as

$$\underline{f} = \sum_{\gamma \in \Gamma} f_\gamma \gamma.$$

In this setting we can add any two sequences $\underline{f}, \underline{g} \in \Sigma(\Gamma, \mathcal{M})$, i.e., $\underline{f} + \underline{g} = \sum (f_\gamma + g_\gamma) \gamma$, and by $\underline{0}$ we denote the sequence with all values equal to 0. For the decimation operators ∂_v^H we have the obvious equation

$$\partial_v^H(\underline{f} + \underline{g}) = \partial_v^H(\underline{f}) + \partial_v^H(\underline{g}).$$

Let $\text{End}(\mathcal{M}) = \{h : \mathcal{M} \to \mathcal{M} \mid h(a + b) = h(a) + h(b) \text{ for all } a, b \in \mathcal{M}\}$ denote the endomorphisms of the monoid $\mathcal{M}$. We have the usual addition, denoted by $+$, and composition, denoted by $\circ$, of endomorphisms. The constant map which maps every $a \in \mathcal{M}$ on 0 is denoted by 0.

The set $\Sigma_c(\Gamma, \text{End}(\mathcal{M})) \subset \Sigma(\Gamma, \text{End}(\mathcal{M}))$ denotes the sequences with finite support.

For $p, q \in \Sigma_c(\Gamma, \text{End}(\mathcal{M}))$ we define the Cauchy product $p * q$ as

$$(p * q)(\gamma) = \sum_{\rho\tau=\gamma} p_\rho \circ q_\tau,$$

where $p = \sum p_\gamma \gamma$ and $q = \sum q_\gamma \gamma$.

We endow $\Sigma_c(\Gamma, \text{End}(\mathcal{M}))$ with a norm $\| \ \|$ defined by

$$\|p\| = \max\{\|\gamma\| \mid \gamma \in \text{supp } p\}$$

for $p \neq 0$ and $\|0\| = 0$. Then we have $\|p + q\| \leq \max\{\|p\|, \|q\|\}$ and $\|p * q\| \leq \|p\| + \|q\|$.

Finally, we define a product $\cdot$ of elements of $\Sigma_c(\Gamma, \text{End}(\mathcal{M}))$ with elements of $\Sigma(\Gamma, \mathcal{M})$ by

$$p \cdot \underline{f} = \sum_{\gamma \in \Gamma} \left(\sum_{\rho\tau=\gamma} p_\rho(f_\tau) \right) \gamma.$$

Lemma 5.1.1.

1. $0 \cdot \underline{f} = \underline{0}$.

2. $p \cdot (\underline{f} + \underline{g}) = p \cdot \underline{f} + p \cdot \underline{g}$.

3. $(p + q) \cdot \underline{f} = p \cdot \underline{f} + q \cdot \underline{f}$.

4. $q \cdot (p \cdot \underline{f}) = (q * p) \cdot \underline{f}$.

5. *Let V be a residue set for the expanding map $H : \Gamma \to \Gamma$ with expansion ratio $C > 1$ and let $r = \max\{\|v\| \mid v \in V\}$. Then for $p \in \Sigma_c(\Gamma, \text{End}(\mathcal{M}))$ we have*

$$\|\partial_v^H(p)\| \leq \frac{\|p\| + r}{C}.$$

6. *Let $p \in \Sigma_c(\Gamma, \text{End}(\mathcal{M}))$ and $\underline{f} \in \Sigma(\Gamma, \mathcal{M})$, then*

$$\partial_v^H(p \cdot H_*(\underline{f})) = \partial_v^H(p) \cdot \underline{f}$$

holds for the v-decimations.

7. $H_*(p \cdot \underline{f}) = H_*(p) \cdot H_*(\underline{f})$.

Proof. The assertions 1, 2, 3, 5, 6, and 7 are obvious. To prove 4, we calculate

$$q \cdot (p \cdot \underline{f})(\gamma) = \sum_{\rho} q_{\rho} \Big(\sum_{\xi} p_{\xi}(f_{\xi^{-1}\rho^{-1}\gamma}) \Big).$$

Since $q_{\rho} \in \text{End}(\mathcal{M})$ and due to the commutativity of $+$ on $\mathcal{M}$ we obtain

$$q \cdot (p \cdot \underline{f})(\gamma) = \sum_{\rho,\xi} q_{\rho} \circ p_{\xi}(f_{\xi^{-1}\rho^{-1}\gamma}),$$

which is the same as $(q * p) \cdot \underline{f}(\gamma)$. $\qquad\square$

We are interested in the invariance of the finiteness of the H-kernel under certain operations of $\Sigma_c(\Gamma, \text{End}(\mathcal{M}))$ on $\Sigma(\Gamma, \mathcal{M})$. We start with a simple observation. We consider an element $h \in \text{End}(\mathcal{M})$ as an element in $\Sigma_c(\Gamma, \text{End}(\mathcal{M}))$ by writing $p = h\,e$, i.e., p is a map from Γ to $\text{End}(\mathcal{M})$ with value h at the neutral element $e \in \Gamma$ and value 0 for all $\gamma \neq e$. Then we have $p \cdot \underline{f} = \sum h(f_{\gamma})\gamma$, i.e., in the notation of the previous chapters, $h \cdot \underline{f} = \hat{h}(\underline{f})$. Thus, we obtain

$$\partial_v^H(h \cdot \underline{f}) = \partial_v^H \hat{h}(\underline{f}) = \hat{h}(\partial_v^H(\underline{f})) = h \cdot \partial_v^H(\underline{f}),$$

cf. 3. of the remark on p. 39. Moreover, we have the trivial statement that the sum of two sequences with finite H-kernel has again a finite H-kernel.

Lemma 5.1.2. *Let $p \in \Sigma_c(\Gamma, \text{End}(\mathcal{M}))$ and let $\underline{f} \in \Sigma(\Gamma, \mathcal{M})$ have a finite H-kernel, then $p \cdot \underline{f}$ has a finite H-kernel.*

Proof. Consider $\xi \in \Gamma$ and the polynomial $p = h\xi \in \Sigma_c(\Gamma, \text{End}(\mathcal{M}))$, where $h \in \text{End}(\mathcal{M}), h \neq 0$. Then

$$p \cdot \underline{f} = \sum_{\gamma} h(f_{\xi^{-1}\gamma}) = h(T_{\xi^{-1}})^*(\underline{f}).$$

By Theorem 3.2.5 and due to the above observation, we conclude that $p \cdot \underline{f}$ has a finite H-kernel.

For an arbitrary $p = \sum p_{\gamma}\,\gamma \in \Sigma_c(\Gamma, \text{End}(\mathcal{M}))$, where $p_{\gamma} = h_{\gamma}\,\gamma$ and $h_{\gamma} \in \text{End}(\mathcal{M})$, we write $p \cdot \underline{f}$ as a finite sum

$$p \cdot \underline{f} = \sum_{\gamma \in \text{supp}(p)} p_{\gamma} \cdot (\underline{f});$$

each of the summands has finite H-kernel and therefore $p \cdot \underline{f}$ has a finite h-kernel. $\square$

We are now prepared to state the main result of this section.

Theorem 5.1.3. *Let $p \in \Sigma_c(\Gamma, \mathrm{End}(\mathcal{M}))$. If $\underline{f} \in \Sigma(\Gamma, \mathcal{M})$ satisfies*

$$\underline{f} = p \cdot H_*(\underline{f}), \tag{5.1}$$

then $\underline{f}$ has a finite H-kernel.

Proof. We shall show that the H-kernel of $\underline{f}$ is finite. For this purpose we fix a residue set V for the expanding map H with expansion ration $C > 1$. By κ we denote the associated image-part-map. With $r = \max\{\|v\| \mid v \in V\}$ and $R = \frac{\|p\|+r}{C-1}$ we set

$$\overline{K} = \{q \cdot \underline{f} \mid q \in \Sigma_c(\Gamma, \mathrm{End}(\mathcal{M})) \text{ such that } \|q\| \leq R\}.$$

Since $\underline{f} \in \overline{K}$ and $\overline{K}$ is a finite set, it remains to show that $\overline{K}$ is invariant under decimation operators. To this end, let $\underline{h} = q \cdot \underline{f}$ be in $\overline{K}$. Since $\underline{f}$ satisfies Equation (5.1) and due to 4. of Lemma 5.1.1, we obtain

$$q \cdot \underline{f} = q \cdot p \cdot H_*(\underline{f}) = (q * p) \cdot H_*(\underline{f}).$$

Due to 6. of Lemma 5.1.1, we obtain that $\partial_v^H(q \cdot \underline{f}) = \partial_v^H(q * p) \cdot \underline{f}$. We therefore estimate

$$\|\partial_v^H(q * p)\| \leq \frac{\|q * p\| + r}{C} \leq \frac{\|q\| + \|p\| + r}{C} \leq R$$

by the choice of R. Therefore $\overline{K}$ is decimation invariant. □

As the next example shows the above proof actually allows a calculation of the kernel of an assumed solution.

Example. Let $\mathcal{M} = \mathbb{F}_2$ be the field with two elements and let $\Gamma = \langle x \rangle = \mathbb{Z}$. Then $H(x^\alpha) = x^{2\alpha}$ is expanding. We fix the residue set as $V = \{x^0, x^1\}$. In this setting, we consider $\mathbb{F}_2$ as a subset of $\mathrm{End}(\mathbb{F}_2)$, where $0 \in \mathbb{F}_2$ is considered as the 0-map in $\mathrm{End}(\mathbb{F}_2)$ and $1 \in \mathbb{F}_2$ is considered as the identity in $\mathrm{End}(\mathbb{F}_2)$. This means that we have the inclusion $\Sigma(\Gamma, \mathbb{F}_2) \subseteq \Sigma(\Gamma, \mathrm{End}(\mathbb{F}_2))$. The multiplication of $p \in \Sigma_c(\Gamma, \mathbb{F}_2)$ is the Cauchy product of a polynomial with a formal Laurent series. The equation

$$\underline{f} = (1 + x + x^2)H_*(\underline{f})$$

has at least one non-trivial solution. We calculate the kernel of the assumed solution as

$$\partial_{x^0}^H(\underline{f}) = (1 + x)\underline{f}$$
$$\partial_{x^1}^H(\underline{f}) = \underline{f}$$
$$\partial_{x^0}^H((1 + x)\underline{f}) = \partial_{x^0}^H((1 + x^3)H_*(\underline{f})) = \underline{f}$$
$$\partial_{x^1}^H((1 + x)\underline{f}) = x\underline{f}$$
$$\partial_{x^0}^H(x\underline{f}) = \partial_{x^0}^H((x + x^2 + x^3)H_*(\underline{f})) = x\underline{f}$$
$$\partial_{x^1}^H(x\underline{f}) = (1 + x)\underline{f}.$$

By the proof of Theorem 2.2.19 there exists a substitution which is associated with the above kernel. In the next section, we shall discuss whether the fixed points of the substitution give a solution of the Mahler equation.

Furthermore, we can also obtain the (reduced) 2-kernel graph of a solution. It is shown in Figure 5.1

Figure 5.1. Reduced kernel graph for Example 1.

Corollary 5.1.4. *Let* $p, q \in \Sigma_c(\Gamma, \mathrm{End}(\mathcal{M}))$. *If* $\underline{f} \in \Sigma(\Gamma, \mathcal{M})$ *satisfies*

$$\underline{f} = q + p \cdot H_*(\underline{f}),$$

then $\underline{f}$ *has a finite H-kernel.*

Proof. Let V be a residue set of H with expansion ratio $C > 1$ and let $r = \max\{\|v\| \mid v \in V\}$, $r^* = \max\{\|\gamma\| \mid \gamma \in \mathrm{supp}\ \underline{q}\}$. Set $R = \frac{\max\{\|\underline{p}\|, r^*\} + r}{C - 1}$. We prove that the set

$$\overline{K} = \{\underline{s} + t \cdot \underline{f} \mid s, t \in \Sigma_c(\Gamma, \mathrm{End}(\mathcal{M})) \text{ such that } \|s\| \leq R \text{ and } \|t\| \leq R\}$$

is invariant under the decimation operators. Since $\underline{f} \in \overline{K}$ and $\overline{K}$ is a finite set, the assertion follows. We have

$$\partial_v^H(s + t \cdot \underline{f}) = \partial_v^H(s + t \cdot \underline{q}) + \partial_v^H(t * p)\underline{f}$$

due to the equation for $\underline{f}$. Applying 5. of Lemma 5.1.1 and the properties of the norm on $\Sigma_c(\Gamma, \mathrm{End}(\mathcal{M}))$, we obtain $\|\partial_v^H(t * p)\| \leq R$. Moreover, we have $\mathrm{supp}(\underline{s} + t \cdot \underline{q}) \subset B_R(e)$. □

More important is the next consequence

Corollary 5.1.5. *Let* $p_j \in \Sigma_c(\Gamma, \mathrm{End}(\mathcal{M}))$, $j = 1, \dots, N$ *and* $q \in \Sigma_c(\Gamma, \mathcal{M})$. *If* $\underline{f}$ *satisfies*

$$\underline{f} = q + p_1 \cdot H_*(\underline{f}) + \cdots + p_N \cdot H_*^N(\underline{f}), \qquad (5.2)$$

then $\underline{f}$ *has a finite H-kernel.*

Proof. Let $\underline{f}_1 = \underline{f}, \underline{f}_2 = H_*(\underline{f}), \ldots, \underline{f}_N = H_*^{N-1}(\underline{f})$. Then Equation (5.2) transforms into

$$\underline{f}_1 = p_1 \cdot H_*(\underline{f}_1) + \cdots + p_N \cdot H_*(\underline{f}_N) + \underline{q}$$
$$\underline{f}_2 = H_*(\underline{f}_1)$$
$$\vdots$$
$$\underline{f}_N = H_*(\underline{f}_{N-1}).$$

Let $\overline{\mathcal{B}} = \mathcal{M}^N$ with distinguished element $(0, \ldots, 0)$ and componentwise addition, i.e., $(a_1, \ldots, a_N) + (b_1, \ldots, b_N) = (a_1 + b_1, \ldots, a_N + b_N)$. Then $\mathrm{End}(\overline{\mathcal{B}})$ is given by $\mathrm{End}(\mathcal{M})^{N \times N}$, i.e., by matrices whose entries are elements of $\mathrm{End}(\mathcal{M})$. For $A = (h_{i,j})_{i,j=1,\ldots,N}$ we define

$$(A(a_1, \ldots, a_N))_k = \sum_{l=1}^{N} h_{k,l}(a_l).$$

With this notation the transformed Equation (5.2) can be interpreted as

$$\underline{F} = P \cdot H_*(\underline{F}) + Q,$$

where $\underline{F} \in \Sigma(\Gamma, \mathcal{M}^N)$ with $\underline{F}(\gamma) = (\underline{f}_1(\gamma), \ldots, \underline{f}_N(\gamma))$, and $Q \in \Sigma_c(\Gamma, \mathcal{M}^N)$ with $Q(\gamma) = (\underline{q}(\gamma), 0, \ldots, 0)$ and $P \in \Sigma_c(\Gamma, \mathrm{End}(\overline{\mathcal{B}}))$ defined by

$$P_\gamma = \begin{pmatrix} (p_1)_\gamma & (p_2)_\gamma & \cdots & (p_N)_\gamma \\ \mathrm{id} & 0 & \cdots & 0 \\ 0 & \mathrm{id} & \cdots & 0 \\ & \vdots & & \\ 0 & \cdots & \mathrm{id} & 0 \end{pmatrix}.$$

Thus, we have shown that $\underline{F}$ satisfies the requirements of Theorem 5.1.3 and Corollary 5.1.5. Therefore, $\underline{F}$ has a finite H-kernel and, by projection, $\underline{f}$ has a finite H-kernel, too. $\square$

Equations of type (5.2) are called Mahler equations. One may hope that Corollaries 5.1.4, 5.1.5 even hold for the case that $\underline{q}$ has a finite H-kernel. Indeed, this is true as we shall show, with different methods, in the next section.

Finally, we arrive at the characterization of H-automatic sequences. A combination of the proofs of Theorem 2.2.19 and Theorem 5.1.3 provides us with the following:

Theorem 5.1.6. *A sequence $\underline{f} \in \Sigma(\Gamma, \mathcal{M})$, where $\mathcal{M}$ is a commutative, associative monoid with 0, is H-automatic if and only if there exists an $N \in \mathbb{N}$, $\overline{\mathcal{B}} = \mathcal{M}^N$, $P \in \Sigma_c(\Gamma, \mathrm{End}(\overline{\mathcal{B}}))$ and $\underline{F} \in \Sigma(\Gamma, \overline{\mathcal{B}})$ such that*

$$\underline{F} = P \cdot H_*(\underline{F})$$

and $\hat{\theta}(\underline{F}) = \underline{f}$, where $\theta : \overline{\mathcal{B}} \to \mathcal{M}$ is defined by $\theta(a_1, \ldots, a_n) = a_1$.

An equation of the type in Theorem 5.1.6 is called an N-dimensional Mahler equation.

5.2 Solutions of Mahler equations

In this section we study Mahler equations over the finite, commutative, and associative monoid $\mathcal{M}$ in more detail. As we have seen in the previous section, any solution $\underline{f}$ of an equation of the type

$$\underline{f} = p \cdot H_*(\underline{f})$$

is H-automatic, or, equivalently, has a finite H-kernel. Since there exists always the trivial solution $\underline{f} = \underline{0}$, we are mainly interested in non-trivial solutions of the above equation. In this section, we develop a method based on substitutions which allows us to find all solutions of the above equations.

We also present a method based on the reduced kernel graph of an assumed solution. This method is less efficient than the method based on substitutions.

We begin with the method based on substitutions. At first glance one might expect that the proof of Theorem 5.1.3 in combination with the proof of Theorem 2.2.19 provides such a tool. The steps for finding a non-trivial solution would then be the following.

Firstly, calculate the H-kernel of an assumed solution. This is basically the proof of Theorem 5.1.3. The H-kernel of $\underline{f}$ and the relations between the elements of the kernel lead to a substitution, as shown in the proof of Theorem 2.2.19.

Secondly, find all fixed points of the associated substitution and check which of the fixed points projects to a solution of the Mahler equation. It is the last step which causes the main difficulties, even if V is a complete digit set of H. The next example shows the drawbacks of the approach.

Examples.

1. Let $\mathcal{M} = \{0, 1\}$ be considered as the field $\mathbb{F}_2$. Let $\Gamma = \langle x \rangle = \{x^l \mid l \in \mathbb{Z}\}$, i.e., Γ is isomorphic to $\mathbb{Z}$. The map $H(x) = x^3$ is an expanding group endomorphism w.r.t. the generating element x. We consider the following equation:

$$\underline{f} = (1 + x + x^3)H_*(\underline{f}).$$

The set $V = \{x^0, x^1, x^2\}$ is a residue set for H. Then the (V, H)-kernel of an assumed solution $\underline{f}$ has four elements, namely

$$\ker_{V,H}(\underline{f}) = \{\underline{f_1} = \underline{f}, \underline{f_2} = \partial_{x^0}^H(\underline{f}), \underline{f_3} = \partial_{x^1}^H(\partial_{x^0}^H(\underline{f})), \underline{f_4} = \partial_{x^2}^H(\underline{f})\}.$$

We thus obtain a substitution $S : \Sigma(\Gamma, \mathbb{F}_2^4) \to \Sigma(\Gamma, \mathbb{F}_2^4)$ defined by

$$s_0(a_1, a_2, a_3, a_4) = (a_2, a_2, a_4, a_4)$$
$$s_1(a_1, a_2, a_3, a_4) = (a_1, a_3, a_2, a_4)$$
$$s_2(a_1, a_2, a_3, a_4) = (a_4, a_1, a_1, a_4).$$

Since $\operatorname{Per}\kappa = \{x^0, x^{-1}\}$ and x^0, x^{-1} are fixed points of κ we can calculate the number of fixed points of the substitution as

$$\left|\{a \in \mathbb{F}_2^4 \mid s_0(a) = a\}\right| \left|\{a \in \overline{\mathscr{A}}^4 \mid s_2(a) = a\}\right|,$$

which is 16. However, as it will turn out, not all fixed points are solutions of the equation.

2. Let the setting be as in 1, except for V. Now let $V_c = \{x^{-1}, x^0, x^1\}$ be a complete digit set. The (V_c, H)-kernel is given by

$$\ker_{V_c, H}(\underline{f}) = \{\underline{f}_1 = \underline{f}, (1 + x)\underline{f}, x\underline{f}\}$$

and the kernel graph is shown in Figure 5.2.

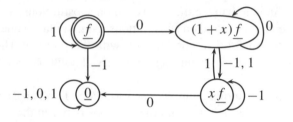

Figure 5.2. (V_c, H)-kernel graph for Example 2.

The substitution is given by

$$s_0(a_1, a_2, a_3, a_4) = (a_2, a_2, a_4, a_4)$$
$$s_1(a_1, a_2, a_3, a_4) = (a_1, a_3, a_3, a_4)$$
$$s_{-1}(a_1, a_2, a_3, a_4) = (a_4, a_4, a_3, a_4).$$

The number of fixed points is equal to 4. However, the number of different solutions of the equation is equal to 2.

Note that the kernel graph together with the output function $\theta : \ker_{V_c, H}(\underline{f}) \to \{0, 1\}$ with $\theta(\underline{g}) = \underline{g}(x^0)$ gives an automaton which generates the two possible solutions.

The examples indicate that the above method is not well suited for finding solutions of Mahler equations. We explain the underlying idea for a better method to solve these equations.

Let $\mathscr{M}$ be the usual monoid and consider the simple Mahler equation

$$\underline{f} = p \cdot H_*(\underline{f}) = \left(\sum_{v \in V} h_v v\right) \cdot H_*(\underline{f}),$$

where V is a residue set of H and $p \in \Sigma(\Gamma, \text{End}(\mathcal{M}))$. The product of p with $H_*(\underline{f})$ is then given by $p \cdot H_*(\underline{f}) = \sum_{v \in V, \gamma \in \Gamma} h_v(f_\gamma) v H(\gamma)$, i.e., the map $\underline{f} \mapsto p \cdot H_*(\underline{f})$ is a substitution. A solution of the simple Mahler equation is a fixed point of the substitution, and vice versa.

For an arbitrary $p \in \Sigma_c(\Gamma, \text{End}(\mathcal{M}))$ and its associated Mahler equation we shall prove a similar result. Namely, we shall show that there exists a finite set $\overline{\mathcal{B}}$, an injective map $\iota : \Sigma(\Gamma, \mathcal{M}) \to \Sigma(\Gamma, \overline{\mathcal{B}})$, and a substitution $S_p : \Sigma(\Gamma, \mathcal{B}) \to \Sigma(\Gamma, \overline{\mathcal{B}})$ such that

$$S_p(\iota(\underline{f})) = \iota(p \cdot H_*(\underline{f}))$$

holds for all $\underline{f} \in \Sigma(\Gamma, \mathcal{M})$. In a dynamical interpretation, we can say that the map $\underline{f} \mapsto p \cdot H_*(\underline{f})$ is conjugate to a substitution S_p on $\iota(\Sigma(\Gamma, \mathcal{M}))$. Moreover, any fixed point of the substitution S_p is an element of $\iota(\Sigma(\Gamma, \mathcal{M}))$ The following example gives a hint for an understanding of the general case.

Example. Let id be the identity on $\mathcal{M} = \{0, 1\}$. We consider the Mahler equation

$$\underline{f} = (\text{id } x^0 + \text{id } x^1 + \text{id } x^3) H_*(\underline{f}),$$

where $H(x) = x^3$. Let $V = \{x^{-1}, x^0, x^1\}$ be a complete digit set of H. We want to find a (non-trivial) solution $\underline{f} \in \Sigma(\mathbb{Z}, \mathcal{M})$ of the above equation. To this end, let $\underline{f} = \sum f_l x^l$. Then we can calculate

$$
\begin{aligned}
(p \cdot H_*(\underline{f}))(x^{3l-2}) &= \underline{f}(x^{l-1}) \\
(p \cdot H_*(\underline{f}))(x^{3l-1}) &= 0 \\
(p \cdot H_*(\underline{f}))(x^{3l}) &= \underline{f}(x^{l-1}) + \underline{f}(x^l) \\
(p \cdot H_*(\underline{f}))(x^{3l+1}) &= \underline{f}(x^l).
\end{aligned}
\tag{5.3}
$$

We can conclude that a knowledge of $\underline{f}(x^l)$ and of $\underline{f}(x^{l-1})$ enables us to compute $(p \cdot H_*(\underline{f}))(x^j)$ for $j = 3l - 2, 3l - 1, 3l, 3l + 1$. If we set $\overline{\mathcal{B}} = \mathcal{M}^2$ and define the injective map $\iota : \Sigma(\Gamma, \mathcal{M}) \to \Sigma(\Gamma, \overline{\mathcal{B}})$ by

$$\iota(\underline{f})(x^l) = \begin{pmatrix} \underline{f}(x^l) \\ \underline{f}(x^{l-1}) \end{pmatrix},$$

then Equation (5.3) defines a substitution $S : \Sigma(\Gamma, \overline{\mathcal{B}}) \to \Sigma(\Gamma, \overline{\mathcal{B}})$ via

$$
\begin{aligned}
s_{-1}(a_1, a_2) &= (0, a_1) \\
s_0(a_1, a_2) &= (a_1 + a_2, 0) \\
s_1(a_1, a_2) &= (a_1, a_1 + a_2).
\end{aligned}
$$

By construction, we have $S(\iota(\underline{f})) = \iota((1 + x + x^3) H_*(\underline{f}))$. Since V is a complete digit set, the number of solutions of the substitution equals the number of fixed points of s_0, which are $(0, 0)$ and $(1, 0)$. In fact, these are all solutions of the Mahler equation. We thus have seen that a sort of grouping of $\underline{f}(x^j)$, where j is in a finite set, gives a new finite set $\overline{\mathcal{B}}$ and new sequences. In this context the operation of p on $\Sigma(\Gamma, \mathcal{M})$ is a substitution. This idea will be our guiding rule for the general situation.

From now on $\Lambda \subset \Gamma$ denotes a finite subset of Γ and $\overline{\mathcal{B}}_\Lambda = \{h : \Lambda \to \mathcal{M}\}$ is the set of maps from Λ to $\mathcal{M}$, where the zero-map, denoted by $0 : \Lambda \to \mathcal{M}, 0(\lambda) = 0$, is the distinguished element.

Definition 5.2.1. Let $\Lambda \subset \Gamma$ be finite. The map $\iota_\Lambda : \Sigma(\Gamma, \mathcal{M}) \to \Sigma(\Gamma, \overline{\mathcal{B}}_\Lambda)$, where $\iota_\Lambda(\underline{f})_\gamma : \Lambda \to \mathcal{M}$ is defined by

$$\iota_\Lambda(\underline{f})_\gamma(\lambda) = \underline{f}(\lambda\gamma),$$

is called Λ-*block-map*. A sequence $\underline{h} \in \Sigma(\Gamma, \overline{\mathcal{B}}_\Lambda)$ is called *admissible* if $\underline{h} \in \iota_\Lambda(\Sigma(\Gamma, \mathcal{M}))$.

Remarks.

1. It is plain that ι_Λ is injective.

2. $\iota_\Lambda(\underline{f}_1 + \underline{f}_2) = \iota_\Lambda(\underline{f}_1) + \iota_\Lambda(\underline{f}_2)$.

3. $\underline{h} = \sum_\gamma h_\gamma \gamma \in \Sigma(\Gamma, \overline{\mathcal{B}}_\Lambda)$ is admissible if and only if $h_\gamma(\lambda) = h_{\gamma'}(\lambda')$ for all $\gamma, \gamma' \in \Gamma, \lambda, \lambda' \in \Lambda$ such that $\lambda\gamma = \lambda'\gamma'$.

There exist several projections from $\Sigma(\Gamma, \overline{\mathcal{B}}_\Lambda)$ to $\Sigma(\Gamma, \mathcal{M})$, some of them are important for us. For $\lambda \in \Lambda$ we define $p_\lambda : \Sigma(\Gamma, \overline{\mathcal{B}}_\Lambda) \to \Sigma(\Gamma, \mathcal{M})$ by

$$p_\lambda\left(\sum h_\gamma \gamma\right) = \sum h_\gamma(\lambda)\gamma$$

and $p_\lambda^* : \Sigma(\Gamma, \overline{\mathcal{B}}_\Lambda) \to \Sigma(\Gamma, \mathcal{M})$ as

$$p_\lambda^*\left(\sum h_\gamma \gamma\right) = \sum h_\gamma(\lambda)\lambda\gamma = (T_\lambda)^* p_\lambda(\underline{h}).$$

The projections p_λ^* allow a characterization of admissible sequences. A sequence $\underline{h} \in \Sigma(\Gamma, \overline{\mathcal{B}}_\Lambda)$ is admissible if and only if $p_\lambda^*(\underline{h}) = p_{\lambda'}^*(\underline{h})$ for all $\lambda, \lambda' \in \Lambda$. Moreover, $p_\lambda^*(\iota_\Lambda(\underline{f})) = \underline{f}$, i.e., p_λ^* is injective on the set of admissible sequences.

The next lemma shows that the map ι_Λ respects the finiteness of H-kernels.

Lemma 5.2.2. *If $\underline{f} \in \Sigma(\Gamma, \mathcal{M})$ has a finite H-kernel, then $\iota_\Lambda(\underline{f})$ has a finite H-kernel, too.*

If $\underline{h} \in \Sigma(\Gamma, \overline{\mathcal{B}}_\Lambda)$ has a finite H-kernel, then $p_\lambda(\underline{h})$ has a finite H-kernel for all $\lambda \in \Lambda$.

Proof. We have the following identity: $p_\lambda \circ \partial_v^H(\underline{h}) = \partial_v^H \circ p_\lambda(\underline{h})$ for all $\underline{h} \in \Sigma(\Gamma, \overline{\mathcal{B}}_\Lambda)$ and $\lambda \in \Lambda$. Thus for $\underline{h} = \iota_\Lambda(\underline{f})$, we conclude that

$$p_\lambda \circ \partial_v^H(\underline{h}) = \partial_v^H \circ (T_\lambda)^*(\underline{f}).$$

Due to Theorem 3.2.5 and due to the finiteness of Λ, we obtain that the H-kernel of $\iota_\Lambda(\underline{f})$ is contained in the set

$$\overline{K} = \left\{ \sum g_\gamma \gamma \mid p_\lambda(\sum g_\gamma \gamma) \in \ker(T_\lambda)^*(\underline{f}) \text{ for all } \lambda \in \Lambda \right\}$$

which is finite.

The second assertion follows from the above commutativity of the projections p_λ and the v-decimations. $\qquad\qquad\qquad\qquad\qquad\qquad\qquad\qquad\qquad\qquad\qquad$ $\square$

Since $p_\lambda^* = (T_\lambda)^* \circ p_\lambda$, we obtain that $p_\lambda^*(\underline{h})$ has a finite H-kernel if $\underline{h}$ has a finite H-kernel.

The following theorem states that for a certain choice of a finite set $\Lambda \subset \Gamma$ and its associated map ι_Λ we can model the operation of $p \in \Sigma_c(\Gamma, \mathrm{End}(\mathcal{M}))$ on elements of the form $H_*(\underline{f})$ by a substitution $S_p : \Sigma(\Gamma, \overline{\mathcal{B}}_\Lambda) \to \Sigma(\Gamma, \overline{\mathcal{B}}_\Lambda)$.

We remind of the notion of the product of two (finite) sets. If $\Lambda_1, \Lambda_2 \subset \Gamma$, then the product $\Lambda_1 \Lambda_2$ is given by $\Lambda_1 \Lambda_2 = \{ \gamma_1 \gamma_2 \mid \gamma_1 \in \Lambda_1, \gamma_2 \in \Lambda_2 \}$.

Theorem 5.2.3. *Let* $p \in \Sigma_c(\Gamma, \mathrm{End}(\mathcal{M}))$ *and let* V *be a fixed residue set of* H. *Then there exists a finite set* $\Lambda = \Lambda(p, V)$ *and a* (V, H)-*substitution* $S_p : \Sigma(\Gamma, \overline{\mathcal{B}}_\Lambda) \to \Sigma(\Gamma, \overline{\mathcal{B}}_\Lambda)$ *such that*

$$\iota_\Lambda(p \cdot H_*(\underline{f})) = S_p(\iota_\Lambda(\underline{f}))$$

holds for all $\underline{f} \in \Sigma(\Gamma, \mathcal{M})$.

Proof. We start with the defining properties of Λ. For $p \in \Sigma_c(\Gamma, \mathrm{End}(\mathcal{M}))$ let the set W be defined by $W = \{ \gamma^{-1} \mid p(\gamma) \neq 0 \}$. In order to describe the operation of p on elements of the form $H_*(\underline{f})$ as a substitution we need a finite set Λ such that

$$W \Lambda v H(\gamma) \cap H(\Gamma) \subset H(\Lambda) H(\gamma)$$

holds for all $\gamma \in \Gamma$ and $v \in V$. In fact, this equation for Λ was solved for the above example. The equation for Λ is equivalent to

$$W \Lambda V \cap H(\Gamma) \subset H(\Lambda).$$

Now observe that for any subset Λ' of Γ the following inclusion is true:

$$\Lambda' \subset \bigcup_{v \in V} v H(\kappa(\Lambda')).$$

Therefore we obtain

$$W \Lambda V \cap H(\Gamma) \subset H(\kappa(W \Lambda V)).$$

Thus we can conclude that a solution Λ of the equation $H(\kappa(W\Lambda V)) = H(\Lambda)$ gives a solution of the above inclusion. If we apply H^{-1}, we obtain: if Λ is a solution of $\kappa(W\Lambda V) = \Lambda$, then Λ satisfies the desired inclusions.

The next step is to show that there exists always a solution Λ. Let $r = \max\{\|v\| \mid v \in V\}$ and $r^* = \max\{\|w\| \mid w \in W\}$, let $R = \frac{2r+r^*}{C-1}$, where C is the expansion ratio of H. If Λ is a non-empty subset of $B_R(e)$, then due to the choice of R we have the following inequality for the norm of $\gamma \in \kappa(W\Lambda V)$:

$$\|\gamma\| = \|\kappa(w\lambda v)\| \leq \frac{r + \|w\lambda v\|}{C}$$
$$\leq \frac{2r + r^* + R}{C}$$
$$\leq R.$$

Therefore, if $\emptyset \neq \Lambda \subset B_R(e)$, then $\kappa(W\Lambda V)$ is a non-empty subset of $B_R(e)$, too. Since the set of subsets of $B_R(e)$ is finite there exists a subset which is a periodic point of the map $\Lambda \mapsto \kappa(W\Lambda V)$. If we take the union over the orbit of the periodic point we obtain a finite set Λ with $\Lambda = \kappa(W\Lambda V)$. We thus have established the existence of a finite set Λ with the desired property.

From now on Λ denotes a solution of $\Lambda = \kappa(W\Lambda V)$. With Λ we are able to define the substitution S_p. An element $\underline{h} \in \overline{\mathcal{B}}_\Lambda$ can also be considered as an element of $\Sigma(\Gamma, \mathcal{M})$. If we consider $\underline{h}$ as an element of $\Sigma(\Gamma, \mathcal{M})$ we write it as $\underline{h}' = \sum_{\lambda \in \Lambda} h(\lambda)\lambda$ with the usual convention $\underline{h}(\gamma) = 0$ for $\gamma \notin \Lambda$.

Due to the choice of Λ we can define maps $s_v : \overline{\mathcal{B}}_\Lambda \to \overline{\mathcal{B}}_\Lambda$ by

$$s_v(h) = \iota_\Lambda(p \cdot H_*(\underline{h}'))(v),$$

where $v \in V$. Since $s_v(0) = 0$ for all $v \in V$, we can define a substitution $S_p : \Sigma(\Gamma, \overline{\mathcal{B}}_\Lambda) \to \Sigma(\Gamma, \overline{\mathcal{B}}_\Lambda)$ by $S_p(\underline{h}) = \sum(T_v \circ H)_* \circ s_v(\underline{h})$.

Finally, we prove that $S_p \circ \iota_\Lambda(\underline{f}) = \iota_\Lambda(p \cdot H_*(\underline{f}))$ holds for all $\underline{f}$.

Let $\lambda, \lambda' \in \Lambda$. We start with

$$\iota_\Lambda(p \cdot H_*(\underline{f}))_{vH(\gamma)}(\lambda') = (p \cdot H_*(\underline{f}))(\lambda'vH(\gamma)) = \sum_{w,\xi} p_{w^{-1}}(f_\xi),$$

where the sum is over all $w \in W$ and $\xi \in \Gamma$ such that $H(\xi) = w\lambda'vH(\gamma)$.

For the substitution we obtain:

$$S_p(\iota_\Lambda(\underline{f}))_{vH(\gamma)}(\lambda') = s_v(\iota_\Lambda(\underline{f})_\gamma)(\lambda')$$
$$= \iota_\Lambda\left(p \cdot H_*\left(\sum_{\lambda \in \Lambda} f_{\lambda\gamma}\lambda\right)\right)_v(\lambda')$$
$$= \left(p \cdot H_*\left(\sum_{\lambda \in \Lambda} f_{\lambda\gamma}\right)\right)(\lambda'v)$$
$$= \sum_{w,\lambda} p_{w^{-1}}(f_{\lambda\gamma}),$$

where the sum is over all $w \in W$ and $\lambda \in \Lambda$ such that $H(\lambda) = w\lambda'v$ holds. Once again, we encounter the defining equation for Λ. The defining equation guarantees that the last sum is in fact equal to the above sum. More formally, we obtain that $H(\xi) = w\lambda'vH(\gamma) = H(\lambda\gamma)$, i.e., $\xi = \lambda\gamma$. Therefore both sums are equal which proves the assertion.   □

We say that the substitution S_p is induced by the equation $\underline{f} = p \cdot H_*(\underline{f})$, or simply is induced by p.

Example. Let $\Gamma = \langle x \rangle$, $H(x^j) = x^{3j}$ and let $V_c = \{x^{-1}, x^0, x^1\}$ be a complete digit set. Consider the Mahler equation

$$\underline{f}(x) = (\mathrm{id} + \mathrm{id}\ x + \mathrm{id}\ x^4) \cdot H_*(\underline{f}),$$

where id is the identity id on $\mathcal{M}$. Then $W = \{x^0, x^{-1}, x^{-4}\}$ and $\Lambda = \{x^{-2}, x^{-1}, x^0\}$ satisfies

$$\kappa(W\Lambda V) = \Lambda.$$

Therefore $\overline{\mathscr{B}} = \mathcal{M}^\Lambda = \mathcal{M}^3 = \{(a_0, a_{-1}, a_{-2}) \mid a_i \in \mathcal{M}, i = 0, -1, -2\}$. The multiplication $(\mathrm{id} + \mathrm{id}\ x + \mathrm{id}\ x^4) \cdot H_*(\underline{f})$ is modelled by the induced substitution $S_p : \Sigma(\Gamma, \mathcal{M}^\Lambda) \to \Sigma(\Gamma, \mathcal{M}^\Lambda)$ defined by

$$s_{-1}(a_0, a_{-1}, a_2) = (0, a_1 + a_{-2}, a_{-1})$$
$$s_0(a_0, a_{-1}, a_{-2}) = (a_0, 0, a_{-1} + a_{-2})$$
$$s_1(a_0, a_{-1}, a_{-2}) = (a_0 + a_{-1}, a_0, 0).$$

The fixed points of the substitution S_p are given by $(a, 0, b)$, where $a, b \in \overline{\mathcal{A}}$. These fixed points give all solutions of the Mahler equation.

If we combine the proofs of Corollary 5.1.5 and Theorem 5.2.3, we obtain a substitution for Mahler equations of the type

$$\underline{f} = p_1 \cdot H_*(\underline{f}) + \cdots + p_N \cdot H_*^{\circ N}(\underline{f}).$$

As a consequence of Theorem 5.2.3 we get an answer to our question on the number of solutions of Mahler equations.

Corollary 5.2.4. *Suppose V is a complete digit set of H and S_p is the induced (V, H)-substitution. Then the number of fixed points of S_p equals the number of solutions of the equation*

$$\underline{f} = p \cdot H_*(\underline{f}).$$

Proof. The proof of Theorem 5.2.3 shows that S_p maps admissible sequences on admissible sequences. It therefore remains to show that any fixed point of S_p is admissible. To this end, suppose that $\underline{h} = \sum h_\gamma \gamma$ is a fixed point of S_p, then $\underline{h}'_e = \sum_{\lambda \in \Lambda} h_e(\lambda)$, $\lambda \in \Sigma(\Gamma, \mathcal{M})$ maps via ι_Λ to an admissible sequence $\iota_\Lambda(\underline{h}'_e)$, which coincides with the assumed fixed point at e. Due to the proof of Lemma 2.2.9, the value at the fixed point of κ completely determines the dynamics. We thus obtain that the sequence $S_p^N(\iota_\Lambda(\underline{h}'_e))$ is a sequence of admissible sequences and converges to the fixed point. Therefore the fixed point is an admissible sequence, and the projection p_λ^* is a solution of the Mahler equation. Since the projection is injective the assertion follows.   □

Remarks.

1. If V is such that Per κ contains only fixed points we get a similar result. Let $\underline{h}$ be a fixed point of S_p. Consider $\underline{g} = \sum_{\xi \in \text{Per}\kappa} h_\xi \xi$, and suppose g satisfies

$$g_\xi(\lambda) = g_{\xi'}(\lambda')$$

whenever $\xi\lambda = \xi'\lambda'$ for all $\xi, \xi' \in \text{Per}\,\kappa$ and $\lambda, \lambda' \in \Lambda$. Then

$$f = \sum_{\xi \in \text{Per}\,\kappa,\ \lambda \in \Lambda} g_\xi(\lambda)\lambda\xi$$

   is mapped under ι_Λ to an admissible sequence which coincides on Per κ with the fixed point $\underline{h}$. The number of solutions of the Mahler equation is then given by the number of fixed points of the substitution with the above property.

2. In the general situation one has that the number of solutions of the Mahler equation is given by the number of admissible fixed points of the associated substitution.

As another application of Theorem 5.2.3 we obtain a generalization of Corollary 5.1.4.

Corollary 5.2.5. *Let $p \in \Sigma_c(\Gamma, \text{End}(\mathcal{M}))$, and let $\underline{g} \in \Sigma(\Gamma, \mathcal{M})$ with a finite H-kernel. If $\underline{f}$ satisfies*

$$\underline{f} = p \cdot H_*(\underline{f}) + \underline{g},$$

then $\underline{f}$ has a finite H-kernel.

Proof. By Theorem 5.2.3 there exist a finite set Λ and a (V, H)-substitution S_p : $\Sigma(\Gamma, \overline{\mathcal{B}}_\Lambda) \to \Sigma(\Gamma, \overline{\mathcal{B}}_\Lambda)$ such that

$$\iota_\Lambda(p \cdot H_*(\underline{f})) = S_p(\iota_\Lambda(\underline{f})).$$

Therefore we consider the Mahler equation over the set $\Sigma(\Gamma, \overline{\mathcal{B}}_\Lambda)$, i.e., we consider the equation

$$\underline{h} = S_p(\underline{h}) + \iota_\Lambda(\underline{g}).$$

By Lemma 5.2.2, $\iota_\Lambda(g)$ has a finite H-kernel. Moreover, an admissible solution $\underline{h}$ projects via $p_\lambda^*(\underline{h})$ to a solution of the Mahler equation. By Lemma 5.2.2, the finiteness of ker $p_\lambda^*(\underline{h})$ follows from the finiteness of the H-kernel of $\underline{h}$. To this end, we define the set $\overline{K}$ by

$$\overline{K} = \{\hat{a}(\underline{h}) + \sum_{\underline{j} \in \ker \iota_\Lambda(g)} \hat{b}_{\underline{j}}(\underline{j}) \mid \text{ where } a, b_{\underline{j}} \in \text{End}(\mathscr{B}_\Lambda) \text{ and } \underline{j} \in \ker \iota_\Lambda(g)\}.$$

The set $\overline{K}$ is finite and contains $\underline{h}$. It is therefore sufficient to show the decimation invariance of $\overline{K}$. To this end, we calculate

$$\partial_v^H(\hat{a}(\underline{h}) + \sum \hat{b}_{\underline{j}}(\underline{j}))$$

$$= \hat{a}(\partial_v^H(\underline{h})) + \sum \hat{b}'_{\underline{j}}(\underline{j})$$

$$= \hat{a}\partial_v^H(S_p(\underline{h}) + \iota_\Lambda(\underline{g})) + \sum \hat{b}'_{\underline{j}}(\underline{j}) \qquad \text{by fixed point property of } \underline{h}$$

$$= \hat{a}(\widehat{s_v}(\underline{h})) + \hat{a}(\partial_v^H(\iota_\Lambda(\underline{g}))) + \sum \hat{b}'_{\underline{j}}(\underline{j}) \quad \text{by Lemma 2.2.17.}$$

The last line shows that $\partial_v^H(\overline{K}) \subset \overline{K}$. By projection, we obtain the desired result for the H-kernel of $p_\lambda^* \underline{h}$. $\qquad \square$

For the number of different solutions of non-homogenous Mahler equations we get a result similar to Corollary 5.2.4.

Corollary 5.2.6. *Let V be a complete digit set and let $g \in \Sigma(\Gamma, \mathcal{M})$ with finite H-kernel. Then the number of solutions of the non-homogenous Mahler equation*

$$\underline{f} = p \cdot H_*(\underline{f}) + \underline{g}$$

is equal to the number of fixed points of the equation

$$\underline{h} = S_p(\underline{h}) + \iota_\Lambda(\underline{g}).$$

Yet another application of Theorem 5.2.3 is concerned with a criterion for a residue set to be a complete digit set.

Corollary 5.2.7. *Let V be a residue set of H, and let $\mathcal{M} = \{0, 1\}$ be considered as the field $\mathbb{F}_2$. Let W be a residue set of H, $W \neq V$. The residue set W is a complete digit set if and only if the equation*

$$\underline{f} = (\sum_{w \in W} \text{id}_{\overline{\mathscr{A}}} w) \cdot H_*(\underline{f}) \qquad (5.4)$$

has two solutions.

Proof. We start with a simple observation. Since W is a residue set of H there are always the solutions $\underline{f} \equiv 0$ and $\underline{f} \equiv 1$ of Equation (5.4).

Now suppose that W is a complete digit set. We apply Theorem 5.2.3 to Equation (5.4) and residue set W and obtain $\Lambda = \{e\}$ and $s_w(a) = a$ for all $w \in W$. Therefore 5.4 has only two solutions.

Suppose that 5.4 has two solutions and that W is not a complete digit set. Then $\Gamma_e = \bigcup_{n=0}^{\infty} \kappa_{H,W}^{-n}(e) \neq \Gamma$ and

$$\underline{f}_e = \sum_{\gamma \in \Gamma_e} \gamma$$

is a solution of 5.4; this is a contradiction.   □

If $\mathcal{M}$ has a richer algebraic structure, e.g., a commutative ring with 0 and 1, we can transfer all the above results to this situation in the following way. The ring multiplication defines a left operation on $\Sigma(\Gamma, \mathcal{M})$ (componentwise multiplication), i.e., we consider $\Sigma(\Gamma, \mathcal{M})$ as an $\mathcal{M}$-module. The set $\Sigma_c(\Gamma, \mathrm{End}(\mathcal{M}))$ is then the set of polynomials with coefficients in $\mathcal{M}$, i.e.,

$$p = \sum_{\gamma \in \Gamma} p_\gamma \gamma$$

with $p_\gamma \in \mathcal{M}$ and $p_\gamma = 0$ almost everywhere. Hence we can identify $\Sigma_c(\Gamma, \mathrm{End}(\mathcal{M}))$ with $\Sigma_c(\Gamma, \mathcal{M}) = \{f \in \Sigma(\Gamma, \mathcal{M}) \mid f \text{ has finite support}\}$. The operation of $p \in \Sigma_c(\Gamma, \overline{\mathcal{A}})$ on $\Sigma(\Gamma, \overline{\mathcal{A}})$ is the usual Cauchy product. If $\mathcal{M}$ is a commutative ring then Lemma 5.1.2 states that the set of H-automatic sequences is a $\Sigma_c(\Gamma, \mathcal{M})$-module.

We conclude this section by comparing the substitution approach for the solutions of a Mahler equation with the approach of using the kernel graph of an assumed solution. As the example above shows, one can determine the solutions of the Mahler equation $\underline{f} = (\mathrm{id} + \mathrm{id}\, x + \mathrm{id}\, x^4) \cdot H_*(\underline{f})$ for every monoid $\mathcal{M}$. An attempt to find the solutions using the kernel graph requires to compute the kernel graph of an assumed solution. It is obvious that this computation depends heavily on the cardinality of $\mathcal{M}$ and also on the structure of the addition in $\mathcal{M}$.

5.3 Γ abelian

In this section, we study the theory of substitutions for the case that Γ is an abelian group. As we have already seen the commutativity of Γ implies that Γ is equal to $\mathbb{Z}^n$ for some $n \in \mathbb{N}$, see Lemma 2.1.3.

We also impose some further restriction on the finite set $\overline{\mathcal{A}}$. Namely we suppose that $\overline{\mathcal{A}}$ carries the structure of a finite field which we denote by $\mathcal{F}$. As before, we also write $\mathbb{F}_q$ for the finite field with $q = p^\alpha$ elements, p a prime number.

The main goal of this chapter is to obtain a sort of inverse statement of Theorem 5.1.3. We shall prove that for $\underline{f} \in \Sigma(\mathbb{Z}^n, \mathcal{F})$, the finiteness of the H-kernel of $\underline{f}$ implies the existence of a generalized Mahler equation for which $\underline{f}$ is a solution. In the course of the proof it will become clear that the conditions that Γ is commutative and that $\mathcal{F}$ is a field are crucial.

From now on we think of $\Gamma = \mathbb{Z}^n$ with generating set $\{x_1, \ldots, x_n\}$. The elements of Γ are written as $x^\alpha = x_1^{\alpha_1} x_2^{\alpha_2} \ldots x_n^{\alpha_n}$ and we have $x^\alpha y^\beta = x_1^{\alpha_1 + \beta_1} \ldots x_n^{\alpha_n + \beta_n}$. Any expanding endomorphism (w.r.t. the generating set $\{x_1, \ldots, x_n\}$) of $\mathbb{Z}^n$ is given by a matrix $M \in \mathbb{Z}^{n \times n}$ which is expanding w.r.t. any norm on $\mathbb{R}^n$. Usually, we use the euclidian norm on $\mathbb{R}^n$.

For an expanding endomorphism H and a residue set V we have the following trivial identities for the associated remainder- and image-part-map, respectively,

$$\kappa(xy) = \kappa(\zeta(x)\zeta(y))\kappa(x)\kappa(y)$$
$$\zeta(xy) = \zeta(\zeta(x)\zeta(y)). \tag{5.5}$$

Similar to Lemma 4.1.6 we obtain a product-formula for decimations.

Lemma 5.3.1. *Let $\mathcal{M}$ be a commutative, finite monoid and let $p \in \Sigma_c(\mathbb{Z}^n, \mathrm{End}(\mathcal{M}))$. Then, for $H : \mathbb{Z}^n \to \mathbb{Z}^n$ expanding,*

$$\partial_v^H(p \cdot \underline{f}) = \sum_{\substack{u, w \in V \\ \zeta(uw) = v}} (T_{\kappa(uw)})_* \circ \partial_u^H(p) \circ \partial_w^H(\underline{f})$$

holds for all $v \in V$, V a residue set of H.

Proof. We have for $x \in \mathbb{Z}^n$ and $v \in V$ that

$$\partial_v^H(p \cdot \underline{f})(x) = (p \cdot \underline{f})(vH(x)) = \sum_{y,z} p_y(f_z),$$

where the sum is over all $y, z \in \mathbb{Z}^n$ such that $yz = vH(x)$. For $y = uH(y')$, $z = wH(z')$ and by Equation (5.5) we get

$$\sum_{y,z} p_y(f_z) = \sum_{\substack{u, w \in V \\ \zeta(uw) = v}} \left(\sum_{y',z'} p_{uH(y')}(f_{wH(z')}) \right),$$

where the last sum is over all y', z' such that $y'z' = \kappa(uw)^{-1}x$. This means that

$$\sum_{y',z'} p_{uH(y')}(f_{wH(z')}) = (T_{\kappa(uw)})_* \circ \partial_u^H(p) \circ \partial_w^H(\underline{f})(x)$$

which proves the assertion.   □

Remarks.

1. The product formula is also valid for the $*$-product of two elements in $\Sigma_c(\mathbb{Z}^n, \text{End}(\mathcal{M}))$.

2. If $\mathcal{M}$ is a commutative ring, then the product formula holds for all $\underline{f}, \underline{g} \in \Sigma(\mathbb{Z}^n, \mathcal{M})$ as long as the products are well defined. E.g., if supp $\underline{f}$, supp $\underline{g} \subset \mathbb{N}^N$ the Cauchy product is well defined and the product formula holds.

After these preliminary remarks we turn our attention to a partial converse of Theorem 5.1.3 and its consequences. Namely, we shall show that for any expanding map $H : \mathbb{Z}^n \to \mathbb{Z}^n$ and any $\underline{f} \in \Sigma(\mathbb{Z}^n, \mathcal{F})$ with finite H-kernel, where $\mathcal{F}$ is a finite field, there exists an $N \in \mathbb{N}$ and polynomials $p_j \in \Sigma_c(\mathbb{Z}^n, \text{End}(\mathcal{F})) = \Sigma_c(\mathbb{Z}^n, \mathcal{F})$, $j = 0, 1, \ldots, N$, such that $\underline{f}$ satisfies a generalized Mahler equation, i.e.,

$$p_0 \underline{f} + p_1 H_*(\underline{f}) + \cdots + p_n H_*^N(\underline{f}) = 0.$$

Before we state two lemmata which are important for a proof of the above statement, we remark that the set of polynomials, i.e., $\Sigma_c(\mathbb{Z}^n, \mathcal{F})$, equipped with Cauchy product and componentwise addition is a commutative integral domain, i.e., a commutative ring without zero divisors. As in Section 5.1, we also work with the sequence space $\Sigma(\mathbb{Z}^n, \mathcal{F}^N)$ and its associated polynomial ring $\Sigma_c(\mathbb{Z}^n, \text{End}(\mathbb{Z}^n, \mathcal{F}^N)) = \Sigma_c(\mathbb{Z}^n, \mathcal{F}^{N \times N})$ which is not commutative. The elements of $\Sigma_c(\mathbb{Z}^n, \mathcal{F}^{N \times N})$ are polynomials with coefficients in the set of $N \times N$-matrices with entries in the field $\mathcal{F}$. These polynomials operate in an obvious way on $\Sigma(\mathbb{Z}^n, \mathcal{F}^N)$.

If $q \in \Sigma_c(\mathbb{Z}^n, \mathcal{F})$ and $\underline{F} \in \Sigma(\mathbb{Z}^n, \mathcal{F}^N)$, then

$$(q\underline{F})(x) = \begin{pmatrix} (q\underline{f}_1)(x) \\ \vdots \\ (q\underline{f}_N)(x) \end{pmatrix}.$$

The following lemma shows that $\underline{f}_1 \in \{\underline{f}_1, \ldots, \underline{f}_N\}$ satisfies a generalized Mahler equation if the elements $\underline{f}_1, \ldots, \underline{f}_N$ are related by a functional equation.

Lemma 5.3.2. *Let $\mathcal{F}$ be a finite field and let $H : \mathbb{Z}^n \to \mathbb{Z}^n$ be expanding. Let $\underline{F} \in \Sigma(\mathbb{Z}^n, \mathcal{F}^N)$ be such that there exist a polynomial $q \in \Sigma_c(\mathbb{Z}^n, \mathcal{F}), q \neq 0$, and $A \in \Sigma_c(\mathbb{Z}^n, \mathcal{F}^{N \times N})$ with*

$$q H_*(\underline{F}) = A\underline{F}. \tag{5.6}$$

Then each component $\underline{f}_j$ of $\underline{F}$ satisfies a generalized Mahler equation.

Proof. It suffices to prove that $\underline{f}_1$ satisfies a generalized Mahler equation. Let $\langle \underline{f}_1, \ldots, \underline{f}_N \rangle$ be the $\Sigma_c(\mathbb{Z}^n, \mathcal{F})$-module generated by $\underline{f}_1, \ldots \underline{f}_N$, i.e.,

$$\langle \underline{f}_1, \ldots, \underline{f}_N \rangle = \{ \underline{f} = \textstyle\sum_{j=1}^{N} p_j \underline{f}_j \mid p_j \in \Sigma_c(\mathbb{Z}^n, \mathcal{F}) \}.$$

Then Equation (5.6) implies that the $\Sigma_c(\mathbb{Z}^n, \mathcal{F})$-module $\langle q H_*(\underline{f}_1), \ldots, q H_*(\underline{f}_N) \rangle$ is a submodule of $\langle \underline{f}_1, \ldots, \underline{f}_N \rangle$. Applying H_* to Equation (5.6) we obtain

$$H_*(q) H_*^2(\underline{F}) = H_*(A) H_*(\underline{F})$$

and after multiplication by q we arrive at

$$q H_*(q) H_*^2(\underline{F}) = H_*(A) q H_*(\underline{F})$$

which yields

$$q H_*(q) H_*^2(\underline{F}) = H_*(A) A \underline{F}.$$

In other words, the $\Sigma_c(\mathbb{Z}^n, \mathcal{F})$-module $\langle q H_*(q) H_*^2(\underline{f}_1), \ldots, q H_*(q) H_*^2(\underline{f}_N) \rangle$ is a submodule of $\langle \underline{f}_1, \ldots, \underline{f}_N \rangle$ and of the module $\langle q H_*(\underline{f}_1), \ldots, q H_*(\underline{f}_N) \rangle$. If we set $q_0 = 1 x^0$ and $q_{n+1} = q H_*(q_n)$ and $B_0 = \mathrm{id}\, x^0$ and $B_{n+1} = H_*(B_n) A$, we obtain that

$$q_n H_*^n(\underline{F}) = B_n \underline{F}$$

holds for all $n \in \mathbb{N}$. We thus have constructed a descending sequence of $\Sigma_c(\mathbb{Z}^n, \mathcal{F})$-modules

$$\langle \underline{f}_1, \ldots, \underline{f}_N \rangle \supset \langle q_1 H_*(\underline{f}_1), \ldots, q_1 H_*(\underline{f}_N) \rangle$$

$$\vdots$$

$$\supset \langle q_N H_*^N(\underline{f}_1), \ldots, q_N H_*^N(\underline{f}_N) \rangle.$$

If we consider the matrices $B_n \in \Sigma(\mathbb{Z}^n, \mathcal{F}^{N \times N})$ as elements of $\Sigma_c(\mathbb{Z}^n, \mathcal{F})^{N \times N}$, i.e., as a polynomial-matrices, then B_n has the following property.

Let $\underline{g} \in \langle q_n H_*^n(\underline{f}_1), \ldots, q_n H_*^n(\underline{f}_n) \rangle$ and write

$$\underline{g} = \sum_{j=1}^{N} r_j q_n H_*^n(\underline{f}_j),$$

where $r_j \in \Sigma_c(\mathbb{Z}^n, \mathcal{F})$. Then

$$B_n \begin{pmatrix} r_1 \\ \vdots \\ r_N \end{pmatrix} = \begin{pmatrix} s_1 \\ \vdots \\ s_N \end{pmatrix}$$

and $\underline{g} = \sum_{j=1}^{N} s_j \underline{f}_j$.

We thus obtained that the $\Sigma_c(\mathbb{Z}^n, \mathcal{F})$-module $\langle \underline{f}_1, q_1 H_*(\underline{f}_1), \ldots, q_N H_*^N(\underline{f}_1) \rangle$ is a submodule of $\langle \underline{f}_1, \ldots, \underline{f}_N \rangle$. From the B_j, $j = 0, \ldots, N$, we can construct an $M \in \Sigma_c(\mathbb{Z}^n, \mathcal{F})^{N \times (N+1)}$ such that for

$$\underline{g} = \sum_{j=0}^N r_j q_j H_*^j(\underline{f}_1) \in \langle \underline{f}_1, q_1 H_*(\underline{f}_1), \ldots, q_N H_*^N(\underline{f}_1) \rangle$$

the polynomials defined by

$$(s_1, \ldots, s_N) = M(r_0, \ldots, r_N)^T$$

give a representation of $\underline{g}$ as $\underline{g} = \sum_{j=1}^N s_j \underline{f}_j$. Since $\Sigma_c(\mathbb{Z}^n, \mathcal{F})$ is an integral domain there exist non-trivial polynomials $p_0, \ldots, p_n$ such that $M(p_0, \ldots, p_N)^T = (0, 0, \ldots, 0)$. Therefore $\underline{g} = \sum_{j=0}^N p_j q_j H_*^j(\underline{f}_1)$ equals zero; which is the desired generalized Mahler equation.   □

The next example shows how the proof proceeds.

Example. Let $\mathcal{F} = \mathbb{F}_2$ and let $\Gamma = \mathbb{Z}$ and $H(x^j) = x^{2j}$, and suppose that $\underline{f}_1, \underline{f}_2$ satisfy

$$(1 + x^2) \begin{pmatrix} H_*(\underline{f}_1) \\ H_*(\underline{f}_2) \end{pmatrix} = \begin{pmatrix} 1 & x \\ x & 1 \end{pmatrix} \begin{pmatrix} \underline{f}_1 \\ \underline{f}_2 \end{pmatrix}.$$

Then we obtain $q_2 = 1 + x^2 + x^4 + x^6$ and

$$B_2 = \begin{pmatrix} 1 + x^3 & x + x^2 \\ x + x^2 & 1 + x^3 \end{pmatrix}$$

which yields

$$M = \begin{pmatrix} 1 & 1 & 1 + x^3 \\ 0 & x & x + x^2 \end{pmatrix}.$$

We thus conclude that $(x + x^3, 1 + x, 1)^T$ is mapped on $(0, 0)$ by M. Therefore the generalized Mahler equation

$$0 = (x + x^3) \underline{f}_1 + (1 + x + x^2 + x^3 + x^4) H_*(\underline{f}_1) + (1 + x^2 + x^4 + x^6) H_*^2(\underline{f}_1)$$

is satisfied by $\underline{f}_1$.

Remark. The proof of Lemma 5.3.2 shows that the condition $H : \mathbb{Z}^n \to \mathbb{Z}^n$ expanding is not necessary.

The next lemma is inspired by the following observation. Suppose that $\underline{f} \in \Sigma(\mathbb{Z}^n, \mathcal{F})$ has a finite (V, H)-kernel $\{\underline{f}_1 = \underline{f}, \underline{f}_2, \ldots, \underline{f}_N\}$. Then there exists an $A \in \Sigma_c(\mathbb{Z}^n, \mathcal{F})^{N \times N}$ such that

$$A = \sum_{v \in V} A_v x^v$$

and each A_ν has 0 or 1 as entries. Moreover A is such that

$$\underline{F} = \begin{pmatrix} \underline{f}_1 \\ \vdots \\ \underline{f}_N \end{pmatrix} = AH_*(\underline{F})$$

holds. This is essentially the second part of the proof of Theorem 2.2.19.

The determinant of A is a polynomial in the variables $x_1, \ldots, x_n$. If $\det A$ is different from the zero polynomial, then we can multiply with the adjoint A^{ad} of A and obtain

$$A^{\mathrm{ad}} \underline{F} = (\det A) H_*(\underline{F}).$$

Then by Lemma 5.3.2, $\underline{f}_1$ satisfies a generalized Mahler equation. Unfortunately, there exist automatic sequences such that $\det A = 0$. The next lemma provides a kind of reduction process to overcome this difficulty.

Lemma 5.3.3. *Let* $\underline{F} \in \Sigma(\mathbb{Z}^n, \mathcal{F}^N) = (\underline{f}_1, \ldots, \underline{f}_N)^T$ *and let* $A \in \Sigma_c(\mathbb{Z}^n, \mathcal{F}^{N \times N})$ *and* $p \in \Sigma_c(\mathbb{Z}^n, \mathcal{F})$, $p \neq 0$, *be such that*

$$p\underline{F} = AH_*(\underline{F})$$

holds. If $\det A = 0$, *then there exists an* $\overline{A} \in \Sigma_c(\mathbb{Z}^n, \mathcal{F}^{(N-1) \times (N-1)})$, *a polynomial* $\overline{p} \in \Sigma_c(\mathbb{Z}^n, \mathcal{F})$ *and* $j \in \{1, 2, \ldots, N\}$ *such that* $\overline{F} = (\underline{f}_1, \ldots, \widehat{\underline{f}_j}, \ldots \underline{f}_N)^T \in$ $\Sigma_c(\mathbb{Z}^n, \mathcal{F}^{N-1})$, *where* $\underline{f}_j$ *is omitted, satisfies*

$$\overline{p}\overline{F} = \overline{A}H_*(\overline{F}).$$

Proof. Since $p\underline{f} = AH_*(\underline{F})$ and $\det A = 0$, we have that $A^{\mathrm{ad}} p\underline{F} = 0$. Since $\Sigma_c(\mathbb{Z}^n, \mathcal{F})$ is an integral domain there exists $\alpha_1, \ldots, \alpha_N \in \Sigma_c(\mathbb{Z}^n, \mathcal{F})$ and $j = 1, \ldots, N$ with $\alpha_j \neq 0$ and

$$\alpha_j p\underline{f}_j = \sum_{l=1, l \neq j}^{N} \alpha_l p\underline{f}_l. \tag{5.7}$$

For all $k \in \{1, \ldots, N\}$ we have

$$p\underline{f}_k = \sum_{l=1}^{N} a_{kl} H_*(\underline{f}_l),$$

where $A = (a_{kl})$, $k, l = 1, \ldots, N$, regarded as an element of $\Sigma_c(\mathbb{Z}^n, \mathcal{F})^{N \times N}$. Multiplication by $H_*(\alpha_j p)$ yields

$$H_*(\alpha_j p) p\underline{f}_k = \sum_{l=1}^{N} a_{kl} H_*(\alpha_j p) H_*(\underline{f}_l).$$

From Equation (5.7) we get

$$H_*(\alpha_j p \underline{f}_j) = \sum_{l=1, l \neq j}^{N} H_*(\alpha_l p \underline{f}_l),$$

and thus

$$H_*(\alpha_j p) p \underline{f}_k = \sum_{l=1, l \neq j}^{N} (a_{kl} H_*(\alpha_j p) + a_{kj} H_*(\alpha_l p)) H_*(\underline{f}_l)$$

for all $l = 1, \ldots, N$ and $l \neq j$. For $\overline{p} = H_*(\alpha_j p) p$ and $\overline{A} = (\overline{a}_{k,l}) = (a_{kl} H_*(\alpha_j p) + a_{kj} H_*(\alpha_l p))$, where $l, k = 1, \ldots, N$ and $l, k \neq j$ the assertion is proved.   □

Example. Let $\mathbb{Z}^n = \langle x \rangle$, i.e., $\mathbb{Z}^n$ is isomorphic to $\mathbb{Z}$, let $H(x^j) = x^{2j}$ and let $\mathcal{F} = \mathbb{F}_2$. We consider

$$\begin{pmatrix} \underline{f}_1 \\ \underline{f}_2 \\ \underline{f}_3 \\ \underline{f}_4 \end{pmatrix} = \begin{pmatrix} x & 1 & 0 & 0 \\ 0 & 0 & 1 & x \\ 0 & 0 & 1+x & 0 \\ 0 & 0 & 0 & 1+x \end{pmatrix} \begin{pmatrix} H_*(\underline{f}_1) \\ H_*(\underline{f}_2) \\ H_*(\underline{f}_3) \\ H_*(\underline{f}_4) \end{pmatrix}$$

which is the defining equation for a 2-automaton that generates the paperfolding sequence. Then $\det A = 0$. Multiplying the above equation by

$$A^{\text{ad}} = \begin{pmatrix} 0 & 1+x^2 & 1+x & x(1+x) \\ 0 & 1+x^2 & 1+x & x(1+x) \\ 0 & 0 & 0 & 0 \\ 0 & 0 & 0 & 0 \end{pmatrix}$$

leads to $(1 + x^2) \underline{f}_2 = (1 + x) \underline{f}_3 + x(1 + x) \underline{f}_4$. This yields

$$(1 + x^2) \begin{pmatrix} \underline{f}_1 \\ \underline{f}_3 \\ \underline{f}_4 \end{pmatrix} = \begin{pmatrix} x(1+x^2) & 1+x^2 & x^2(1+x^2) \\ 0 & 1+x^3 & 0 \\ 0 & 0 & 1+x^3 \end{pmatrix} \begin{pmatrix} H_*(\underline{f}_1) \\ H_*(\underline{f}_3) \\ H_*(\underline{f}_4) \end{pmatrix},$$

and the above matrix has a non-vanishing determinant. We can thus multiply by the adjoint and apply Lemma 5.3.2.

With these two lemmata we are able to prove the main result of this chapter.

Theorem 5.3.4. *Let $\underline{f} \in \Sigma(\mathbb{Z}^n, \mathcal{F})$, where $\mathcal{F}$ be is finite field. If $\underline{f}$ has a finite H-kernel, then $\underline{f}$ satisfies a generalized Mahler equation.*

Proof. Let $\ker_{H,V} = \{\underline{f}_1 = \underline{f}, \ldots, \underline{f}_N\}$ be the kernel of $\underline{f}$ and V be a residue set of H. Then there exists a matrix $A \in \Sigma_c(\mathbb{Z}^n, \mathcal{F})^{N \times N}$ such that

$$\underline{F} = \begin{pmatrix} \underline{f}_1 \\ \vdots \\ \underline{f}_N \end{pmatrix} A H_*(\underline{F}).$$

If $\det A \neq 0$, we multiply the above equation by the adjoint A^{ad} of A and obtain

$$A^{\text{ad}} \underline{F} = (\det A) H_*(\underline{F}).$$

By Lemma 5.3.2, $\underline{f}_1$ satisfies a generalized Mahler equation.

If $\det A = 0$, then there exist $\alpha_1, \ldots, \alpha_N \in \Sigma_c(\mathbb{Z}^n, \mathcal{F})$ such that $0 = \alpha_1 \underline{f}_1 + \cdots + \alpha_N \underline{f}_N$ and not all α_j equal zero. We distinguish two cases.

If there is a $j \in \{2, 3, \ldots, N\}$ such that $\alpha_j \neq 0$ we apply Lemma 5.3.3 and obtain a $p_1 \in \Sigma_c(\mathbb{Z}^n, \mathcal{F})$ and $A_1 \in \Sigma_c(\mathbb{Z}^n, \mathcal{F}^{(N-1) \times (N-1)})$ such that for the modified kernel of $\underline{f}$, i.e., the set $\{\underline{f}_i \mid i = 1, \ldots, N, i \neq j\}$ the equation $p_1 \underline{F}_1 = A_1 H_*(\underline{F}_1)$ holds.

If $\alpha_1 \neq 0$ and $\alpha_2 = \alpha_3 = \cdots = \alpha_N = 0$ then $0 = \alpha_1 \underline{f}_1 = \alpha_1 \underline{f}$ is the desired Mahler equation.

Iterating this process we either arrive at an A_j with $\det A_j \neq 0$ and, after multiplication by the adjoint of A_j, we apply Lemma 5.3.2 or we get a Mahler equation of the type $0 = p \underline{f}$, where $p \in \Sigma_c(\mathbb{Z}^n, \mathcal{F})$. $\qquad\square$

The question on the converse of Theorem 5.3.4 is a very delicate one. Its answer depends on the field $\mathcal{F}$ and the expanding map H. Even in the one-dimensional case, i.e., $\Gamma = \mathbb{Z}$, interesting phenomena occur.

Example. Let $\Gamma = \mathbb{Z}$ and $\mathcal{F} = \mathbb{F}_p$, where p is a prime number, the field with p elements, the expanding map $H : \mathbb{Z}^n \to \mathbb{Z}^n$ is defined by $H(x) = x^2$.

We consider the following generalized Mahler equation

$$(1 - x)\underline{f} = H_*(\underline{f}).$$

Then we have the following results.

a) For all prime numbers p there exists a non-trivial solution $\underline{g}_p$ with supp $\underline{g}_p \subset \mathbb{N}$.

b) The solution $\underline{g}_p$ has a finite H-kernel if and only if $p = 2$.

The first statement follows from the fact that the infinite product

$$g(x) = \prod_{j=0}^{\infty} \frac{1}{1 - x^{2^j}} = \prod_{j=0}^{\infty} \left(\sum_{k=0}^{\infty} x^{2^j k} \right)$$

is well defined over $\mathbb{Z}$ and not zero. Moreover, $\underline{g}_p \equiv g(x) \bmod p$ satisfies the above generalized Mahler equation.

For the second assertion, we fix the residue set as $V = \{x^0, x^1\}$ and a prime number $p \geq 3$ and show that the orbit of $\underline{g}_p$ under iteration of $\partial_{x^0}^H$ is not eventually periodic. Therefore the 2-kernel of $\underline{g}_p$ is infinite. Since $\operatorname{supp} \underline{g}_p \subset \mathbb{N}_0$ we can divide by $1 - x$ and thus obtains that $\underline{g}_p$ satisfies

$$\underline{g}_p = \frac{1}{1 - x} H_*(\underline{g}_p). \tag{5.8}$$

As a next step we use Corollary 5.3.1 to calculate the decimations of rational functions. A rational function $\underline{r}$ is of the form $\underline{r} = \underline{p}/\underline{q}$, where $\underline{p}, \underline{q}$ are polynomials.

We have

$$\partial_{x^0}^H\left(\underline{p}/\underline{q}\right) = \frac{\partial_{x^0}^H(\underline{p})\partial_{x^0}^H(\underline{q}) - x\partial_{x^1}^H(\underline{p})\partial_{x^1}^H(\underline{q})}{\partial_{x^0}^H(\underline{q})^2 - x\partial_{x^1}^H(\underline{q})^2}$$

and

$$\partial_{x^1}^H\left(\underline{p}/\underline{q}\right) = \frac{\partial_{x^0}^H(\underline{q})\partial_{x^1}^H(\underline{p}) - \partial_{x^0}^H(\underline{p})\partial_{x^1}^H(\underline{q})}{\partial_{x^0}^H(\underline{q})^2 - x\partial_{x^1}^H(\underline{q})^2}.$$

If we consider $1 - x$ as an element of the polynomial ring $\mathbb{Z}[x]$ then a simple induction argument shows that

$$(1 - x)^n = \partial_{x^0}^H\left((1 - x)^n\right)^2 - x\partial_{x^1}^H\left((1 - x)^n\right)^2$$

holds for all $n \in \mathbb{N}$. Moreover, we have

$$\partial_{x^0}^H\left((1 - x)^n\right)(1) = 2^{n-1} \quad \text{and} \quad \partial_{x^1}^H\left((1 - x)^n\right)(1) = -2^{n-1}$$

for all $n \in \mathbb{N}$.

The product formula and the formula for the decimations of rational functions applied to Equation (5.8) gives for the n-th iterate of $\partial_{x^0}^H$

$$\left(\partial_{x^0}^H\right)^n (\underline{g}_p) = \frac{q_n}{(1 - x)^n} \underline{g}_p,$$

and the recursion

$$q_{n+1}(x) = \partial_{x^0}^H(q_n)\partial_{x^0}^H\left((1 - x)^{n+1}\right) - x\partial_{x^1}^H(q_n)\partial_{x^1}^H\left((1 - x)^{n+1}\right).$$

Combining all these facts we see that $q_{n+1}(1) = 2^n q_0(1)$.

Since $q_n(1)$ is always a power of 2 the polynomial $q_n(x)$ viewed as an element of $\mathbb{Z}_p[x]$ (p prime number greater than 2) has no factor of the form $1 - x$. Therefore all $\left(\partial_{x^0}^H\right)^n (\underline{g}_p)$ are different, i.e., the 2-kernel is infinite. For $p = 2$ it is easy to verify, using the above formulas, that the 2-kernel is finite.

If $\mathcal{F} = \mathbb{F}_{p^\alpha}$ is a finite field of characteristic p, then $\underline{f}$ being a solution of a generalized Mahler equation implies that $\underline{f}$ is automatic.

Theorem 5.3.5. *Let* $\Gamma = \mathbb{Z}$ *and let* $\mathbb{F}_{p^{\alpha}}$ *be a finite field of characteristic* p *and consider the expanding map* $H(x^j) = x^{pj}$. *If* $\underline{f} = \sum_{l \in \mathbb{Z}} x^l f_l \in \Sigma(\mathbb{F}_{p^{\alpha}})$ *is such that there exists an* $m_0 \in \mathbb{Z}$ *with* $f_m = 0$ *for all* $m < m_0$ *and if* $\underline{f}$ *satisfies a generalized Mahler equation, i.e.,*

$$\sum_{j=0}^{N} \underline{p}_j H_*^{\circ j}(\underline{f}) = 0$$

for polynomials $\underline{p}_j \in \Sigma_c(\mathbb{Z}, \overline{\mathcal{A}})$ *not all of them equal to zero, then* $\underline{f}$ *has a finite* H-*kernel.*

Proof. Since $\mathbb{F}_{p^{\alpha}}$ is a finite field of characteristic p, the p^{α}-th power of $\underline{p} \in \Sigma_c(\mathbb{Z}, \overline{\mathcal{A}})$ is given as the α-th iterate of H_*, i.e.,

$$\underline{p}^{p^{\alpha}} = H_*^{\circ \alpha}(\underline{p}). \tag{5.9}$$

It is no restriction to assume that $\underline{p}_0$ is not the zero polynomial. Indeed, suppose that there exists $0 < j_0$ such that $\underline{p}_{j_0}$ is the smallest non-zero polynomial, then there exists at least one decimation ∂_i^H, $i \in \{0, \ldots, p-1\}$, such that $\partial_i^H(\underline{p}_{j_0}) \neq \underline{0}$. This gives

$$0 = \partial_i^H \left(\sum_{j=j_0}^{N} \underline{p}_j H_*^{\circ j}(\underline{f}) = 0 \right) = \sum_{j=j_0-1}^{N} \underline{q}_j H_*^{\circ j}(\underline{f}).$$

Therefore

$$\underline{p}_0 \underline{f} + \underline{p}_1 H_*(\underline{f}) + \cdots + \underline{p}_N H_*(\underline{f}) = 0$$

and $\underline{p}_0$ is different from the zero sequence. Since $\underline{p}_j$ are polynomials and $f_m = 0$ for all m sufficiently small, we may divide by $\underline{p}_0$. If we set $\underline{g}_i = H_*^{\circ(i-1)}(\underline{f})$ for $i = 1, \ldots, N$, then we can transform the above equation into a matrix version:

$$\underline{G} = \begin{pmatrix} \underline{g}_1 \\ \underline{g}_2 \\ \vdots \\ \underline{g}_N \end{pmatrix} = \frac{1}{\underline{p}_0} \begin{pmatrix} -\underline{p}_1 & -\underline{p}_2 & \cdots & -\underline{p}_N \\ 1 & 0 & \cdots & 0 \\ 0 & 1 & \cdots & 0 \\ \vdots & 0 & \ddots & \vdots \\ 0 & 0 & \cdots & 1 \end{pmatrix} \begin{pmatrix} H_*(\underline{g}_1) \\ H_*(\underline{g}_2) \\ \vdots \\ H_*(\underline{g}_N) \end{pmatrix} = Q' H_*(\underline{G}).$$

For α as in Equation (5.9) one obtains a polynomial $\underline{q} \in \Sigma(\mathbb{Z}, \overline{\mathcal{A}})$ and a polynomial matrix $\tilde{Q} \in \Sigma_c(\mathbb{Z}, \overline{\mathcal{A}}^{N \times N})$ such that

$$\underline{G} = \frac{1}{\underline{q}} \tilde{Q} H_*^{\alpha}(\underline{G}).$$

To complete the proof we prove that $\underline{G}$ has a finite H^{α}-kernel. To this end, we consider the set

$$K = \left\{ \frac{1}{\underline{q}} Q \underline{G} \mid Q \in \Sigma(\mathbb{Z}, \overline{\mathcal{A}}_c^{N \times N}), \|Q\| \leq \|\underline{q}\| + \|\tilde{Q}\| + r \right\},$$

where $r = \max\{|v| \mid v \in V\}$ for a residue set V of H^α. Let $v \in V$, then we have

$$\partial_v^{H^\alpha}(\underline{G}) = \partial_v^{H^\alpha}\left(\frac{1}{\underline{q}}\tilde{Q}H_*^\alpha(\underline{G})\right).$$

By Equation (5.9), this can be transformed to

$$\frac{1}{\underline{q}}\partial_v^{H^\alpha}\left(\underline{q}^{p^\alpha-1}\tilde{Q}\right)\underline{G}$$

and

$$\left\|\partial_v^{H^\alpha}\left(\underline{q}^{p^\alpha-1}\tilde{Q}\right)\right\| \leq \frac{(p^\alpha - 1)\|\underline{q}\| + \|\tilde{Q}\| + r}{p^\alpha - 1} \leq \|\underline{q}\| + \|\tilde{Q}\| + r$$

which shows that $\partial_v^{H^\alpha}(\underline{G}) \in K$ for all $v \in V$. Similar arguments show that K is invariant under decimations. This shows that $\underline{G}$ is p^α-automatic and therefore f is p-automatic. □

As an immediate consequence we also note

Corollary 5.3.6. *Let $\mathbb{F}_q$ be a finite field of characteristic p. If $\underline{f} = \sum_{l \in \mathbb{Z}} f_l\, x^l$ is such that there exists an $l_0 \in \mathbb{Z}$ with $f_l = 0$ for all $l \leq l_0$ and if there exist polynomials $\underline{p}_j \in \Sigma_c(\mathbb{Z}, \mathbb{F}_q)$, $j = 0, \dots, N$, such that*

$$\sum_{j=0}^{N} \underline{p}_j \underline{f}^j = 0,$$

then $\underline{f}$ is p-automatic.

Proof. If $\underline{f}$ satisfies the above equation then there exist polynomials $\underline{q}_j \in \Sigma(\mathbb{Z}, \mathbb{F}_q)$, $j = 0, \dots, M$, such that at least one of the polynomials is different from zero and such that

$$\sum_{j=0}^{M} \underline{q}_j H_*^{\circ j}(\underline{f}) = 0$$

holds. Then the assertion follows from Theorem 5.3.5. □

We conclude with the higher dimensional analogue of Theorem 5.3.5.

Theorem 5.3.7. *Let $\underline{f} \in \Sigma(\mathbb{Z}^n, \mathbb{F}_{p^\alpha})$ be such that there exists a $t \in \mathbb{N}^n$ with $t + \mathrm{supp}(\underline{f}) \subset \mathbb{N}^n$. If there exist polynomials $\underline{p}_j$, $j = 0, \dots, N$, not all of them zero, and*

$$\sum_{j=0}^{N} \underline{p}_j H_*^{\circ j}(\underline{f}) = 0,$$

then $\underline{f}$ is H-automatic, where $H(x_1^{a_1} \dots x_n^{a_n}) = x_1^{pa_1} \dots x_n^{pa_n}$.

The proof follows the same lines as the proof of Theorem 5.3.5. Observe that the condition on the support of $\underline{f}$ allows us to multiply $\underline{f}$ with $1/p$ for a polynomial $\underline{p}$.

Examples.

1. A rational function $\underline{f} \in \Sigma(\mathbb{Z}^n, \overline{\mathcal{A}})$ is the quotient of two polynomials $\underline{p}, \underline{q} \in \Sigma(\mathbb{Z}^n, \mathcal{F})$, where $\mathcal{F}$ is a finite field. Therefore, $\underline{f}\underline{q} = \underline{p}$ is an equation for $\underline{f}$. By Corollary 5.3.6, $\underline{f}$ is H-automatic, where $H : \mathbb{Z}^n \to \mathbb{Z}^n$, $H(x) = x^p$.

2. Let $\Gamma = \mathbb{Z}^2$ and $H(x_1^\alpha x_2^\beta) = x_1^{2\alpha} x_2^{3\beta}$. Then $\underline{r} = \sum_{k=0}^\infty x_1^k x_2^k \in \Sigma(\mathbb{Z}^2, \mathbb{Z}_2)$ is a rational function. However, $\underline{r}$ has an infinite H-kernel. This follows from the fact that $\operatorname{supp}(\underline{r})$ is the graph of the identity on $\mathbb{N}$ and an application of Theorem 4.4.5

5.4 Notes and comments

The book [111] provides an excellent source of information on finite fields.

Theorems 5.3.4, 5.3.5 and 5.3.7, and Corollary 5.3.6 are generalizations of results stated in [53], [54], [148], and [149].

The theory of Mahler equations dates back to 1929, see [121]. Mahler equations play an important role in transcendence question, see e.g., [134]. In [117], [118], [119], [120], N-dimensional Mahler equations are considered.

The inverse problem, namely the question whether a solution of a Mahler equation is automatic has been discussed under various aspects in [33], [34], [35], [72], [73], and [99].

Automaticity properties of rational functions in more than one variable are discussed in [94] and [149].

In [147] one finds a treatment of formal power series in non-commuting variables.

Bibliography

[1] Allouche, Jean-Paul, Nouveaux résultats de transcendance de réels à développement non aléatoire. Gaz. Math. 84 (2000), 19–34.

[2] Allouche, Jean-Paul, Algebraic and analytic randomness. In: Noise, oscillators and algebraic randomness (Chapelle des Bois, 1999), Lecture Notes in Phys. 550, Springer-Verlag, Berlin, 2000, 345–356.

[3] Allouche, Jean-Paul, Shallit, Jeffrey, The ubiquitous Prouhet-Thue–Morse sequence. In: Sequences and their applications (Singapore, 1998), Springer Ser. Discrete Math. Theor. Comput. Sci., Springer-Verlag, London, 1999, 1–16.

[4] Allouche, J.-P., Peyrière, J., Wen, Z.-X., Wen, Z.-Y., Hankel determinants of the Thue–Morse sequence. Ann. Inst. Fourier (Grenoble) 48 (1998), 1–27.

[5] Allouche, J.-P., v. Haeseler, F., Lange, E., Petersen, A., Skordev, G., Linear cellular automata and automatic sequences. Cellular automata (Gießen, 1996), Parallel Comput. 23 (1997), 1577–1592.

[6] Allouche, J.-P., v. Haeseler, F., Peitgen, H.-O., Petersen, A., Skordev, G., Automaticity of double sequences generated by one-dimensional linear cellular automata. Theoret. Comput. Sci. 188 (1–2) (1997), 195–209.

[7] Allouche, J.-P., Cateland, E., Peitgen, H.-O., Skordev, G., Shallit, J., Automatic maps on a semiring with digits. Symposium in Honor of Benoit Mandelbrot (Curaçao, 1995), Fractals 3 (1995), 663–677.

[8] Allouche, J.-P., Schrödinger operators with Rudin-Shapiro potentials are not palindromic. Quantum problems in condensed matter physics. J. Math. Phys. 38 (1997), 1843–1848.

[9] Allouche, J.-P., Cateland, E., Gilbert, W. J., Peitgen, H.-O., Shallit, J. O., Skordev, G., Automatic maps in exotic numeration systems. Theory Comput. Syst. 30 (1997), 285–331.

[10] Allouche, J.-P., Mendès France, M., Automata and automatic sequences. In: Beyond quasicrystals (Les Houches, 1994), Springer-Verlag, Berlin, 1995, 293–367.

[11] Allouche, J.-P., v. Haeseler, F., Peitgen, H.-O., Skordev, G., Linear cellular automata, finite automata and Pascal's triangle. Discrete Appl. Math. 66 (1996), 1–22.

[12] Allouche, Jean-Paul, Bousquet-Mélou, Mireille, Facteurs des suites de Rudin-Shapiro généralisées. Journées Montoises (Mons, 1992), Bull. Belg. Math. Soc. Simon Stevin 1 (1994), 145–164.

[13] Allouche, Jean-Paul, Arnold, André, Berstel, Jean, Brlek, Srečko, Jockusch, William, Plouffe, Simon, Sagan, Bruce E., A relative of the Thue–Morse sequence. Formal power series and algebraic combinatorics (Montreal, PQ, 1992), Discrete Math. 139 (1–3) (1995), 455–461.

[14] Allouche, Jean-Paul, Bousquet-Mélou, Mireille, Canonical positions for the factors in paperfolding sequences. Theoret. Comput. Sci. 129 (1994), 263–278.

[15] Allouche, Jean-Paul, q-regular sequences and other generalizations of q-automatic sequences. In: LATIN '92 (São Paulo, 1992), Lecture Notes in Comput. Sci. 583, Springer-Verlag, Berlin, 1992, 15–23.

[16] Allouche, Jean-Paul, Salon, Olivier, Sous-suites polynomiales de certaines suites automatiques. J. Théor. Nombres Bordeaux 5 (1993), 111–121.

[17] Allouche, Jean-Paul, Shallit, Jeffrey. The ring of k-regular sequences, Theoret. Comput. Sci. 98 (1992), 163–197.

[18] Allouche, Jean-Paul, Morton, Patrick, Shallit, Jeffrey, Pattern spectra, substring enumeration, and automatic sequences. Discrete mathematics and applications to computer science (Marseille, 1989), Theoret. Comput. Sci. 94 (1992), 161–174.

[19] Allouche, Jean-Paul, The number of factors in a paperfolding sequence. Bull. Austral. Math. Soc. 46 (1992), 23–32.

[20] Allouche, J.-P., Automates finis, arithmétique et systèmes dynamiques. Rev. Mat. Apl. 9 (1988), 1–8.

[21] Allouche, Jean-Paul, Automates finis en théorie des nombres. Exposition. Math. 5 (1987), 239–266.

[22] Allouche, Jean-Paul, Mendès France, Michel, Quasicrystal Ising chain and automata theory. J. Statist. Phys. 42 (5–6) (1986), 809–821.

[23] Allouche, Jean-Paul, Automates finis et séries de Dirichlet. Seminar on number theory, 1984–1985 (Talence, 1984/1985), Exp. No. 8, Univ. Bordeaux I, Talence, 1985.

[24] Allouche, Jean-Paul, Cosnard, Michel, Une propriété extrémale de suites automatiques. In: Hubert Delange colloquium (Orsay, 1982), Publ. Math. Orsay 83-4, Univ. Paris XI, Orsay, 1983, 1–7.

[25] Allouche, Jean-Paul, Cosnard, Michel, Itérations de fonctions unimodales et suites engendrées par automates. C. R. Acad. Sci. Paris Sér. I Math. 296 (1983), 159–162.

[26] Allouche, J.-P., Cosnard, M., Une propriété extrémale de la suite de Thue–Morse en rapport avec les cascades de Feigenbaum. Seminar on Number Theory, 1981/1982, Exp. No. 26, Univ. Bordeaux I, Talence, 1982.

[27] Arnold, A., Brlek, S., Optimal word chains for the Thue–Morse word. Inform. and Comput. 83 (1989), 140–151.

[28] Axel, Francoise, Peyriére, Jacques, États étendus dans une chaîne à désordre contrôlé. C. R. Acad. Sci. Paris Sér. II Méc. Phys. Chim. Sci. Univers Sci. Terre 306 (1988), 179–182.

[29] Barbé, André, Peitgen, Heinz-Otto, Skordev, Gencho, Automaticity of coarse-graining invariant orbits of one-dimensional linear cellular automata. Internat. J. Bifur. Chaos Appl. Sci. Engrg. 9 (1999), 67–95.

[30] Barbé, André, Complex order from disorder and from simple order in coarse-graining invariant orbits of certain two-dimensional linear cellular automata. Internat. J. Bifur. Chaos Appl. Sci. Engrg. 7 (1997), 1451–1496.

[31] Barbé, A., v. Haeseler, F. , Peitgen, H.-O., Skordev, G., Coarse-graining invariant patterns of one-dimensional two-state linear cellular automata. Internat. J. Bifur. Chaos Appl. Sci. Engrg. 5 (1995), 1611–1631.

[32] Baum, Leonard E., Sweet, Melvin M., Continued fractions of algebraic power series in characteristic 2. Ann. of Math. (2) 103 (1976), 593–610.

[33] Becker, Paul-Georg, k-regular power series and Mahler-type functional equations. J. Number Theory 49 (1994), 269–286.

[34] Becker, Paul-Georg, Transcendence measures for the values of generalized Mahler functions in arbitrary characteristic. Publ. Math. Debrecen 45 (1994), 269–282.

[35] Becker, Paul-Georg, Automatische Folgen und Transzendenz in positiver Charakteristik. Arch. Math. (Basel) 61 (1993), 68–74.

[36] Bellissard, J., Spectral properties of Schrödinger's operator with a Thue–Morse potential. In: Number theory and physics (Les Houches, 1989), Springer Proc. Phys. 47, Springer-Verlag, Berlin, 1990, 140–150.

[37] Bercoff, Christiane, Uniform tag systems for paperfolding sequences. Discrete Appl. Math. 77 (1997), 119–138.

[38] Berthé, Valérie, Sequences of low complexity: automatic and Sturmian sequences. In: Topics in symbolic dynamics and applications (Temuco, 1997), London Math. Soc. Lecture Note Ser. 279, Cambridge University Press, Cambridge, 2000, 1–34.

[39] Berthé, Valérie, Conditional entropy of some automatic sequences. J. Phys. A 27 (1994), 7993–8006.

[40] Berthé, Valérie, De nouvelles preuves "automatiques" de transcendance pour la fonction zêta de Carlitz. Journées Arithmétiques 1991 (Geneva), Astérisque No. 209 (13), (1992), 159–168.

[41] Berstel, J., Crochemore, M., Pin, J.-E., Thue–Morse sequence and p-adic topology for the free monoid. Discrete Math. 76 (1989), 89–94.

[42] Berstel, Jean. Transductions and context-free languages. Leitfäden Angew. Math. Mech. 38, B. G. Teubner, Stuttgart, 1979.

[43] Berstel, Jean, Reutenauer, Christophe, Rational series and their languages. Monogr. Theoret. Comput. Sci. EATCS Ser. 12, Springer-Verlag, Berlin, 1988.

[44] Blanchard, François, Cellular automata and transducers. A topological view. In: Cellular automata, dynamical systems and neural networks (Santiago,1992), Math. Appl. 282, Kluwer Acad. Publ., Dordrecht, 1994, 1–22.

[45] Brauer, Wilfried, Automatentheorie. Eine Einführung in die Theorie endlicher Automaten. Leitfäden Angew. Math. Mech. , B. G. Teubner, Stuttgart, 1984.

[46] Bruyère, Véronique, Hansel, Georges, Michaux, Christian, Villemaire, Roger, Correction to: "Logic and p-recognizable sets of integers". Bull. Belg. Math. Soc. Simon Stevin 1 (1994), 577.

[47] Bruyère, Véronique, Hansel, Georges, Michaux, Christian, Villemaire, Roger, Logic and p-recognizable sets of integers. Journées Montoises (Mons, 1992), Bull. Belg. Math. Soc. Simon Stevin 1 (1994), 191–238.

[48] Büchi, J. Richard, Finite automata, their algebras and grammars. Towards a theory of formal expressions. Edited and with a preface by Dirk Siefkes. Springer-Verlag, New York, 1989.

[49] Buser, Peter, Karcher, Hermann, Gromov's almost flat manifolds. Astérisque 81, Société Mathématique de France, Paris, 1981.

[50] Černý, Anton, On sequences resulting from iteration of modified quadratic and palindromic mappings. Theoret. Comput. Sci. 188 (1997), 161–174.

[51] Černý, Anton, On generalized words of Thue–Morse. Acta Math. Univ. Comenian. 48 (1986), 299–309.

[52] Černý, A., On generalized words of Thue–Morse. In: Mathematical foundations of computer science (Prague, 1984), Lecture Notes in Comput. Sci. 176, Springer-Verlag, Berlin, 1984, 232–239.

[53] Christol, G., Kamae, T., Mendès France, M., Rauzy, G., Suites algébriques, automates et substitutions. Bull. Soc. Math. France 108 (1980), 401–419.

[54] Christol, Gilles, Ensembles presque periodiques k-reconnaissables. Theoret. Comput. Sci. 9 (1979), 141–145.

[55] Cobham, Alan, Uniform tag sequences. Math. Systems Theory 6 (1972), 164–192.

[56] Cobham, Alan, On the base-dependence of sets of numbers recognizable by finite automata. Math. Systems Theory 3 (1969), 186–192.

[57] Combescure, M., Recurrent versus diffusive dynamics for a kicked quantum system. J. Statist. Phys. 62 (1991), 779–791.

[58] Coornaert, Michel, Papadopoulos, Athanase, Symbolic dynamics and hyperbolic groups. Lecture Notes in Math. 1539. Springer-Verlag, Berlin, 1993.

[59] Culik, Karel, II, Friš, Ivan, Weighted finite transducers in image processing. Discrete Appl. Math. 58 (1995), 223–237.

[60] Dassow, Jürgen, Păun, Gheorghe, Regulated rewriting in formal language theory. Monogr. Theoret. Comput. Sci. EATCS Ser. 18, Springer-Verlag, Berlin, 1989

[61] Davis, C., Knuth,D., Number representations and dragon curves, I, II. J. Recreational Math. 3 (1970), 61–81; 133–149.

[62] Dekking, F. M., On the Thue–Morse measure. Acta Univ. Carolin. Math. Phys. 33 (2) (1992), 35–40.

[63] Dekking, F. M., Marches automatiques. J. Théor. Nombres Bordeaux 5 (1993), 93–100.

[64] Dekking, F. M., Iteration of maps by an automaton. Discrete Math. 126 (1994), 81–86.

[65] Dekking, Michel, Mendès France, Michel, van der Poorten, Alf, Folds. III. More morphisms. Math. Intelligencer 4 (1982), 190–195.

[66] Dekking, Michel, Mendès France, Michel, van der Poorten, Alf, Folds. II. Symmetry disturbed. Math. Intelligencer 4 (1982), 173–181.

[67] Dekking, Michel, Mendès France, Michel, van der Poorten, Alf, Folds. Math. Intelligencer 4 (1982), 130–138.

[68] Dekking, Michel, Transcendance du nombre de Thue–Morse. C. R. Acad. Sci. Paris Sér. A-B 285 (1977), A157–A160.

[69] Delyon, F., Peyrière, J., Recurrence of the eigenstates of a Schrödinger operator with automatic potential. J. Statist. Phys. 64 (1991), 363–368.

[70] de Luca, Aldo, Varricchio, Stefano, Some combinatorial properties of the Thue–Morse sequence and a problem in semigroups. Theoret. Comput. Sci. 63 (1989), 333–348.

[71] Doche, Christophe, Mendès France, Michel, Integral geometry and real zeros of Thue–Morse polynomials. Experiment. Math. 9 (2000), 339–350.

[72] Dumas, Philippe, Récurrences mahlériennes, suites automatiques, études asymptotiques. Thèse, Université de Bordeaux I, Talence, 1993. Institut National de Recherche en Informatique et en Automatique (INRIA), Rocquencourt, 1993.

[73] Dumas, Ph., Algebraic aspects of B-regular series. In: Automata, languages and programming (Lund, 1993), Lecture Notes in Comput. Sci. 700, Springer-Verlag, Berlin, 1993, 457–468.

[74] Dumont, Jean-Marie, Thomas, Alain, Modifications de nombres normaux par des transducteurs. Acta Arith. 68 (1994), 153–170.

[75] Drmota, Michael, Skałba, Mariusz, Rarified sums of the Thue–Morse sequence. Trans. Amer. Math. Soc. 352 (2000), 609–642.

[76] Drmota, Michael, Skałba, Mariusz, Sign-changes of the Thue–Morse fractal function and Dirichlet L-series. Manuscripta Math. 86 (1995), 519–541.

[77] Eilenberg, Samuel, Automata, languages, and machines. Vol. B, Pure Appl. Math. 59, Academic Press, New York–London, 1976.

[78] Eilenberg, Samuel, Automata, languages, and machines. Vol. A. Pure Appl. Math. 58, Academic Press, New York, 1974.

[79] Fabre, Stéphane, Substitutions et β-systèmes de numération. Theoret. Comput. Sci. 137 (1995), 219–236.

[80] Fabre, Stéphane, Une généralisation du théorème de Cobham. Acta Arith. 67 (1994), 197–208.

[81] Ferenczi, Sébastien, Tiling the Morse sequence. Discrete mathematics and applications to computer science (Marseille, 1989), Theoret. Comput. Sci. 94 (1992), 215–221.

[82] Frougny, Christiane, Systèmes de numération linéaires et θ-représentations. Discrete mathematics and applications to computer science (Marseille, 1989), Theoret. Comput. Sci. 94 (1992), 223–236.

[83] Fülöp, Zoltán, Vogler, Heiko, Syntax-directed semantics. Formal models based on tree transducers. Monogr. Theoret. Comput. Sci. EATCS Ser., Springer-Verlag, Berlin, 1998.

[84] Gantmacher, Felix R., The theory of matrices. Vol. 1. Translated from the Russian by K. A. Hirsch. Reprint of the 1959 translation. AMS Chelsea Publishing, Providence, RI, 1998.

[85] Ginsburg, Seymour, Algebraic and automata-theoretic properties of formal languages. Fund. Stud. Comput. Sci. 2, North-Holland Publishing Co., Amsterdam–Oxford; American Elsevier Publishing Co., Inc., New York, 1975.

[86] Goldstein, Sheldon, Kelly, Kevin A., Speer, Eugene R., The fractal structure of rarefied sums of the Thue–Morse sequence. J. Number Theory 42 (1992), 1–19.

[87] Goldstein, S., Kelly, K., Lebowitz, J. L., Szasz, D., Asymmetric random walk on a random Thue–Morse lattice. Fractals in physics (Vence, 1989), Phys. D 38 (1989), 141–153.

[88] Gromov, Mikhael, Groups of polynomial growth and expanding maps. Inst. Hautes Études Sci. Publ. Math. 53 (1981), 53–73.

[89] v. Haeseler, Fritz, Jürgensen, Wibke, Irreducible polynomials generated by decimations. In: Finite fields and applications (Augsburg, 1999), Springer-Verlag, Berlin, 2001, 224–231.

[90] v. Haeseler, F., Peitgen, H.-O., Skordev, G., Self-similar structure of rescaled evolution sets of cellular automata. II. Internat. J. Bifur. Chaos Appl. Sci. Engrg. 11 (2001), 927–941.

[91] v. Haeseler, F., Peitgen, H.-O., Skordev, G., Self-similar structure of rescaled evolution sets of cellular automata. I. Internat. J. Bifur. Chaos Appl. Sci. Engrg. 11 (2001), 913–926.

[92] v. Haeseler, F., Peitgen, H.-O., Skordev, G., On the fractal structure of the rescaled evolution set of Carlitz sequences of polynomials. Discrete Appl. Math. 103 (2000), 89–109.

[93] v. Haeseler, Fritz, Jürgensen, Wibke, Automaticity of solutions of Mahler equations. In: Sequences and their applications (Singapore, 1998), Springer Ser. Discrete Math. Theor. Comput. Sci., Springer-Verlag, London, 1999, 228–239.

[94] v. Haeseler, F., Petersen, A., Automaticity of rational functions. Beiträge Algebra Geom. 39 (1998), 219–229.

[95] v. Haeseler, F., On algebraic properties of sequences generated by substitutions over a group. Habilitationsschrift, University of Bremen, 1996.

[96] v. Haeseler, Fritz, Peitgen, Heinz-Otto, Skordev, Gencho, Pascal's triangle, dynamical systems and attractors. Ergodic Theory Dynam. Systems 12 (1992), 479–486.

[97] Hedlund, G. A., Endormorphisms and automorphisms of the shift dynamical system. Math. Systems Theory 3 (1969), 320–375.

[98] Hopcroft, John E., Ullman, Jeffrey D., Introduction to automata theory, languages, and computation. Addison-Wesley Series in Computer Science, Addison-Wesley Publishing Co., Reading, Mass., 1979.

[99] Jürgensen, Wibke, Equation Kernels as a Link between Automaticity and Mahler Equations. PhD thesis, University of Bremen, Shaker Verlag, Aachen, 2000.

[100] Justin, Jacques, Pirillo, Giuseppe, Decimations and Sturmian words. RAIRO Inform. Théor. Appl. 31 (1997), 271–290.

[101] Kamae, T., Mendès France, M., A continuous family of automata: the Ising automata. Ann. Inst. H. Poincaré Phys. Théor. 64 (1996), 349–372.

[102] Karamanos, K., Entropy analysis of substitutive sequences revisited. J. Phys. A 34 (2001), 9231–9241.

[103] Kátai, I., Szabó, J., Canonical number systems for complex integers, Acta Sci. Math. (Szeged) 37 (1975), 255–280.

[104] Keane, M., Generalized Morse sequences, Z. Wahrsch. Verw. Gebiete 10 (1968), 335–353,

[105] Kelley, Dean, Automata and formal languages. An introduction. Prentice Hall, Inc., Englewood Cliffs, NJ, 1995.

[106] Ketkar, Pallavi, Zamboni, Luca Q., Primitive substitutive numbers are closed under rational multiplication. J. Théor. Nombres Bordeaux 10 (1998), 315–320.

[107] Knuth, D., The Art of Computer Programming, vol.2, Addison-Wesley, Reading, MA, 1979.

[108] Koskas, Michel. About the p-paperfolding words. Theoret. Comput. Sci. 158 (1996), 35–51.

[109] Lehr, Siegfried, Shallit, Jeffrey, Tromp, John, On the vector space of the automatic reals. Theoret. Comput. Sci. 163 (1996), 193–210.

[110] Lehr, S., A result about languages concerning paperfolding sequences. Math. Systems Theory 25 (1992), 309–313.

[111] Lidl, L., Niederreiter, H., Finite Fields, Encyclopedia Math. Appl. 20, Addison-Wesley, 1983.

[112] Lin, Zhi Fang, Tao, Rui Bao, Phase transition of quantum Ising spin models on g-letter generalized Thue–Morse aperiodic chains. J. Phys. A 25 (1992), 2483–2488.

[113] Lin, Zhi Fang, Tao, Rui Bao, Quantum Ising model on Thue–Morse aperiodic chain. Phys. Lett. A 150 (1990), 11–13.

[114] Lind, D., Dynamical properties of quasihyperbolic toral automorphisms. Ergodic Theory Dynam. Systems 2 (1982), 42–68.

[115] Lothaire, M., Combinatorics on words. Cambridge University Press, Cambridge, 1997.

[116] Litow, B., Dumas, Ph., Additive cellular automata and algebraic series. Theoret. Comput. Sci. 119 (1993), 345–354.

[117] Loxton, J. H., Automata and transcendence. In: New advances in transcendence theory (Durham, 1986), Cambridge University Press, Cambridge, 1988, 215–228.

[118] Loxton, J. H., van der Poorten, A. J., Arithmetic properties of automata: regular sequences, J. Reine Angew. Math. 392 (1988), 57–69.

[119] Loxton, J. H., van der Poorten, A. J., Arithmetic properties of the solutions of a class of functional equations, J. Reine Angew. Math. 330 (1982), 159–172.

[120] Loxton, J. H., Van der Poorten, A. J., On algebraic functions satisfying a class of functional equations, Aequationes Math. 14 (1976), 413–420.

[121] Mahler, Kurt, Arithmetische Eigenschaften der Lösungen einer Klasse von Funktionalgleichungen. Math. Ann. 101 (1929), 342–366.

[122] Mauduit, Christian, Multiplicative properties of the Thue–Morse sequence. Period. Math. Hungar. 43 (2001), 137–153.

[123] Macdonald, I. D., The Theory of Groups. Oxford University Press, Oxford, 1968.

[124] Meduna, Alexander, Automata and languages. Theory and applications. Springer-Verlag, London, 2000.

[125] Mendès France, Michel, Sebbar, Ahmed, Pliages de papiers, fonctions thêta et méthode du cercle. Acta Math. 183 (1999), 101–139.

[126] Mendès France, M., The inhomogeneous Ising chain and paperfolding. In: Number theory and physics (Les Houches, 1989), Springer Proc. Phys. 47, Springer-Verlag, Berlin, 1990, 195–202.

[127] Mendès France, Michel, The Ising transducer. Ann. Inst. H. Poincaré Phys. Théor. 52 (1990), 259–265.

[128] Mendès France, Michel, Chaos implies confusion. In: Number theory and dynamical systems (York, 1987), London Math. Soc. Lecture Note Ser. 134, Cambridge University Press, Cambridge, 1989, 137–152.

[129] Mendès France, Michel, van der Poorten, Alfred J., Arithmetic and analytical properties of paperfolding sequences, Bull. Austr. Math. Soc., 24 (1981), 123–131.

[130] Milnor, John, Growth of finitely generated solvable groups. J. Differential Geom. 2 (1968), 447–449.

[131] Morse, M., Recurrent geodesics on a surface of negative curvature. Trans. Amer. Math. Soc. 2 (1921), 84–100.

[132] Morton, Patrick, Connections between binary patterns and paperfolding. Sém. Théor. Nombres Bordeaux (2) 2 (1990), 1–12.

[133] Mossé, Brigitte, Reconnaissabilité des substitutions et complexité des suites automatiques. Bull. Soc. Math. France 124 (1996), 329–346.

[134] Nishioka, Kumiko, Mahler functions and transcendence. Lecture Notes in Math. 1631, Springer-Verlag, Berlin, 1996.

[135] Peitgen, Heinz-Otto, Jürgens, Hartmut, Saupe, Dietmar, Chaos and Fractals. Springer-Verlag, New York, 1992.

[136] Petersen, Antje, Automatic Sequences, Rational Functions, and Geometry. PhD thesis, University of Bremen, 1997.

[137] Point, F., Bruyère, V., On the Cobham-Semenov theorem. Theory Comput. Syst. 30 (1997), 197–220.

[138] Prusinkiewicz, Przemysław, Hanan, James, Lindenmayer systems, fractals, and plants. Lecture Notes in Biomath. 79. Springer-Verlag, New York, 1989.

[139] Puchta, Jan-Christoph, Spilker, Jürgen, Die Thue–Morse-Folge. Elem. Math. 55 (2000), 110–122.

[140] Queffélec, M., Transcendance des fractions continues de Thue–Morse. J. Number Theory 73 (1998), 201–211.

[141] Queffélec, Martine, Spectral study of automatic and substitutive sequences. In: Beyond quasicrystals (Les Houches, 1994), Springer-Verlag, Berlin, 1995, 369–414.

[142] Queffélec, Martine, Substitution dynamical systems – spectral analysis. Lecture Notes in Math. 1294, Springer-Verlag, Berlin, 1987.

[143] Rauzy, Gerard, Low complexity and geometry. In: Dynamics of complex interacting systems (Santiago, 1994), Nonlinear Phenom. Complex Systems 2, Kluwer Acad. Publ., Dordrecht, 1996, 147–177.

[144] Razafy Andriamampianina, D., Suites de Toeplitz, p-pliage, suites automatiques et polynômes. Acta Math. Hungar. 73 (1996), 179–190.

[145] Razafy Andriamampianina, D., Le p-pliage de papier et les polynômes. C. R. Acad. Sci. Paris Sér. I Math. 314 (1992), 875–878.

[146] Rodenhausen, Anna, Paperfolding, generalized Rudin-Shapiro sequences, and the Thue–Morse sequence. Symposium in Honor of Benoit Mandelbrot (Curaçao, 1995), Fractals 3 (1995), 679–688.

[147] Salomaa, Arto, Soittola, Matti, Automata-theoretic aspects of formal power series. Texts Monogr. Comput. Sci., Springer-Verlag, New York–Heidelberg, 1978.

[148] Salon, Olivier, Suites automatiques à multi-indices et algébricité. C. R. Acad. Sci. Paris Sér. I Math. 305 (1987), 501–504.

[149] Salon, Oliver, Suites automatiques à multi-indices, Séminaire de Théorie des Nombres de Bordeaux (1986–1987), exposé no. 4.

[150] Séébold, Patrice, Generalized Thue–Morse sequences. Fundamentals of computation theory (Cottbus, 1985), Lecture Notes in Comput. Sci. 199, Springer-Verlag, Berlin, 1985, 402–411.

[151] Seneta, E., Non-negative matrices. An introduction to theory and applications. George Allen & Unwin Ltd., London, 1973.

[152] Shallit, Jeffrey, Number theory and formal languages. Emerging applications of number theory (Minneapolis, MN, 1996), IMA Vol. Math. Appl. 109, Springer-Verlag, New York, 1999, 547–570.

[153] Shallit, Jeffrey, A generalization of automatic sequences. Theoret. Comput. Sci. 61 (1988), 1–16.

[154] Shub, Michael, Endomorphismen of compact differentiable manifolds. Amer. J. Math. 91 (1969), 175–199.

[155] Stanley, Richard P., Enumerative Combinatorics. Vol. 1. Cambridge Stud. Adv. Math. 49, Cambridge University Press, Cambridge, 1997

[156] Straubing, Howard, Finite automata, formal logic, and circuit complexity. Progr. Theoret. Comput. Sci. , Birkhäuser Boston, Inc., Boston, MA, 1994.

[157] Tapsoba, Théodore, Suites automatiques et papiers pliés. IMHOTEP J. Afr. Math. Pures Appl. 1 (1997), 31–40.

[158] Thue, Axel, Über unendliche Zeichenreihen. Christiana Vidensk. Selsk. Skr. 7 (1906), 1–22

[159] Trakhtenbrot, B. A., Barzdinprime, Ya. M., Finite automata. Behavior and synthesis. Fund. Stud. Comput. Sci. 1, North-Holland Publishing Co., Amsterdam–London; American Elsevier Publishing Co., Inc., New York, 1973.

[160] Uchida, Yoshihisa, On p and q-additive functions. Tokyo J. Math. 22 (1999), 83–97.

[161] Verger-Gaugry, Jean-Louis, Mathematical quasicrystals with toric internal spaces. In: Self-similar systems (Dubna, 1998), Joint Inst. Nuclear Res., Dubna, 1999, 260–270.

[162] Verger-Gaugry, Jean-Louis, Wolny, Janusz, Generalized Meyer sets and Thue–Morse quasicrystals with toric internal spaces. J. Phys. A 32 (1999), 6445–6460.

[163] Wen, Zhi Xiong, Wen, Zhi Ying, Marches sur les arbres homogènes suivant une suite substitutive. Sém. Théor. Nombres Bordeaux (2) 4 (1992), 155–186.

[164] Wen, Zhi Xiong, Wen, Zhi Ying, Some studies on the (p, q)-type sequences. Discrete mathematics and applications to computer science (Marseille, 1989), Theoret. Comput. Sci. 94 (1992), 373–393.

[165] Wood, Derick, Theory of computation. Harper & Row Computer Science and Technology Series. Harper & Row, Publishers, New York, 1987.

[166] Wolf, Joseph A., Growth of finitely generated solvable group and curvature of Riemannian manifolds. J. Differential Geom. 2 (1968), 421–446.

[167] Yao, Jia-yan, Généralisations de la suite de Thue–Morse. Ann. Sci. Math. Québec 21 (1997), 177–189.

[168] Zaks, Michael A., Pikovsky, Arkady S., Kurths, Jürgen, On the generalized dimensions for the Fourier spectrum of the Thue–Morse sequence. J. Phys. A 32 (1999), 1523–1530.

Index